What is the World

What is the World

STOYAN KURTEV

Copyright © 2013 by Stoyan Kurtev.

| ISBN: | Softcover | 978-1-4836-8332-4 |
| | eBook | 978-1-4836-8333-1 |

All rights reserved. No part of this book may be reproduced or transmitted in any form or by any means, electronic or mechanical, including photocopying, recording, or by any information storage and retrieval system, without permission in writing from the copyright owner.

Print information available on the last page.

Rev. date: 08/22/2019

To order additional copies of this book, contact:
Xlibris
1-888-795-4274
www.Xlibris.com
Orders@Xlibris.com
599985

Contents

Chapter 1: What is Matter 11
   The Classical Picture: Particles in Space 11
   The Double-Slit Experiment 15
   Particles as Representations 20
   Physical Space and State Space 21
   What is Mass 28
   What are Forces 34
   What is Energy 45
   What is Time 50
   Quantization of Properties 53
   Sums over Histories and Virtual Particles 60
   Wavefunctions and the Schrödinger Equation 64
   What is Spin 70
   What is Entanglement of Particles' States 75
   Heisenberg's Uncertainty Principle 81
   The Double-Slit Experiment—Revisited 85
   Aggregations of Particles (How To Imagine Bulk Matter) 90
   Aggregate States of Matter: Gas, Liquid and Solid 94
   Biological Matter (What is Life) 98
   Evolution as Accumulation of Representations 102
   The Brain as the Culmination of the Evolutionary Process 107

Chapter 2: What is the Brain 109
   Basic facts about the brain 109
   What is Consciousness 111
   How to Imagine Mental States 118
   The Basic Thought Process (Reasoning and Concepts) 121
   What is Language 125
   What is Memory 129
   What is the Unconscious 134
   What is Sleep 137

What are the Senses..140
    Vision 143
    Hearing 148
    Other Senses...150
What are the Emotions 151
What are Mental Illnesses..155
Aging of the Brain
    (Developments Throughout the Course of Life) 159
Is There Free Will...164
The Space of Mental States...167

Chapter 3: What is the Universe 171
Basic Cosmology 171
What is the Big Bang..175
The Dimensionality of the Universe 179
Why is the Universe So Big 186
The Future 192
    500 years into the future 192
    1000 years into the future 194
    5000 years into the future 195
    10 000 years into the future 197
    50 000 years into the future 198
    100 000 years into the future 200
    500 000 years into the future 202
    1 million years into the future...204
    10 million years into the future...206
    Beyond 10 million years into the future 208
The Representational Picture of the World
    (How to Imagine the Universe) 210

Chapter 4: What Does It All Mean 215
What is the Meaning of Life..215
The Role of Knowledge (How to Think Correctly) 218
What is Right and What is Wrong 222
The Present State of Human Thought..............................227
Politics and Economics..232
The Role of Science..242
The Near Future (The Next 50 Years) 246
Conclusion (What Does It All Mean From a Personal Perspective).... 254

Preface

This is a different book. Unlike others, it offers answers to the basic philosophical questions based on a multitude of new insights and realizations, totally transforming the traditional conceptions in the relevant subjects. The key underlying idea is that the phenomenology of matter, as described by modern physics, can be recast in terms of a system of interacting representations. This is the subject of Chapter 1, where we see how the various physical phenomena described by quantum mechanics and relativity can be reformulated as natural outcomes of the behavior of interacting representations.

The picture of interacting representations is a conceptual one. It does not translate readily into mathematical formalism, but on the other hand the elementary particles do not calculate integrals in order to determine how to behave, so the conceptual understanding of matter is as vital for our understanding of the world as the ability to calculate and predict precisely the behavior of particles. The two domains of knowledge complement each other, rather than compete with each other. This book tries to fill the gap in the conceptual understanding of matter and mind, and the way we imagine things when we are not busy calculating.

The new understanding of matter allows us to understand in turn what is special about the human brain as a physical system and how it engenders consciousness. This explanation of consciousness is radically different from the existing theories. Although based on quantum theory and the idea of entanglement, it posits a different mechanism for realizing entanglement at the macroscopic scale in the brain, and in addition, a different role for entanglement in the explanatory framework of the consciousness phenomenon. The workings of the brain and the new explanation of consciousness are the subject of Chapter 2.

Chapter 3 builds on the new picture of matter and consciousness derived from the notion of representations and considers the totality of everything, which we call 'the universe'. It offers reinterpretation of some of the fundamental ideas in cosmology, such as the origin of space and time, the dimensionality of space, dark matter and dark energy, determinism vs. randomness, etc., in light of the new understanding of matter developed in the preceding chapters, and based on these new explanations formulates a prediction for the future evolutionary process of conscious matter and how it can affect the entire universe.

The first three chapters form the main body of the book and tell a self-sufficient story of how we can understand the world starting from the fundamental idea that it consists entirely and exclusively of representations. Chapter 4 is in its character an addendum to the first three, offering the author's personal views on the consequences of the newly developed understanding of the world for our present-day knowledge and the main topics of interest for the general public. Although it does not deal with scientific knowledge, it is nonetheless of equal importance to the first three because of its more pragmatic implications.

Given the novelty of the majority of ideas appearing in the book it would be unreasonable to expect that all of them will turn out to be true, however, the ideas are all related and form one organic whole, thus making it plausible that the majority of them would be either true or false as an integrated set. On the other hand, there is currently no alternative theory with the same breadth and scope of explanatory power that would offer competing explanations, so we have to assume tentatively that the proposed view of the world might be true. Of course, no knowledge is eternal, so it is reasonable to assume that there is an even more compelling answer to the question of what is the world, and the current theory should be viewed only as a step in that direction.

The style of writing is at an intermediate level between light popular-science readings and more demanding academic texts, with the aim to appeal to a broader audience while retaining as much as possible the logical rigor and accuracy of the arguments. This makes the book a not very fluent reading, and may necessitate re-reading of some passages and pauses for reflection, but the end result of understanding what it says should feel rewarding to the reader. There are many brief excursions into more specific topics, especially in the later chapters, which offer food for thought and offer the chance to the reader to extend the arguments in various directions.

As a whole, the book requires intensive use of one's imagination, which is what acquiring new knowledge is all about. Ideally, the explanations offered here would have been laid out in a greater detail to make them easier to comprehend, but this would have made the text much more voluminous and would have brought it to the level of an undergraduate textbook. This may be the aim of a subsequent project. As a first attempt at this novel type of a comprehensive theory of reality, this book ought to serve its purpose rather well.

Stoyan Kurtev
Leicester, UK, June 2012

Chapter 1: What is Matter

The Classical Picture: Particles in Space

The classical picture of matter is that of particles, usually depicted as tiny balls, moving around in empty space and bouncing off each other. Conceptually, this presupposes the existence of empty space and some kind of granular 'stuff' situated in this space, with the ability to move around and interact when two pieces of this stuff come into contact by virtue of occupying adjacent locations in space.

This picture is derived from our everyday experience of the macroscopic world around us, but modern physics has shown that it is inadequate for understanding how things work at the microscopic scale of the elementary particles making up every object in the universe. The theory of relativity stipulates that there is a deep and inextricable link between material objects and the space they populate, while particle physics envisions the vacuum, i.e., empty space, as a kind of primordial foam out of which virtual particles can spring into existence and then vanish again in tiny fractions of a second.

The development of physics through measurements of phenomena in the microscopic world and the corresponding development of the mathematical apparatus describing these phenomena in a formal language necessitated a radical departure from the classical picture of the physical world, forcing physicists to develop quite a different conceptual framework for imagining physical events from the one we are used to from our everyday experience. In the new framework, the existence of space depends on the existence of the material structures inside it, the mass of an object depends on the speed it travels with, there is a limit to how fast an object can travel in space, and there are also the weird quantum effects, like a particle being simultaneously at two different spatial locations, particles overcoming energy barriers by borrowing energy from the vacuum for extremely brief moments, particles transforming into one another in high-energy collisions, energy being transformed into mass and vice versa, etc. All this weirdness came unexpectedly from a historical perspective, and very few inroads have been made into answering the question of *why* all these phenomena are the way they are, i.e., why the world has to function in this way and not according to our classical intuitions.

Although modern physics seems like a radical departure from our common sense grounded in everyday experience, it still has not been able to craft its own logic based on a single unifying idea from which all other physical theories can be derived. Physicists have been in search for Grand Unifying Theories for quite some time, but still without a complete success. In pursuit of the Theory Of Everything they try to study (both theoretically and experimentally) ever smaller spatial scales and ever higher energies, leading to the formulation of theories like supersymmetry and string theory, but many of the theoretical postulates cannot be verified by present-day technology because they concern spatial and energy scales far beyond what is possibly achievable.

The drive towards more fundamental theories has been guided by the tool of choice for physicists—mathematical equations. The ability to calculate precisely the outcome of measurements based on theoretical assumptions forms the core of the work of physicists, while interpretation of the calculations in terms of a conceptual framework explaining what 'really happens' physically comes second, if at all. In some cases, the conceptualization of the physical phenomena described mathematically has proven so hard that some physicists have given up the effort to understand what is going on out there in reality, assuming an attitude which was dubbed by some as "shut up and calculate!" Consequently, the mathematical constructs get more and more complex, and with the development of physical theories more and more experimentally observed phenomena get modeled by mathematical constructs, forming the impression that the physical picture of the world gets more and more complex as we probe deeper into smaller scales when trying to decompose material objects into their constituents. This contradicts common sense; the building blocks of something complex should be simpler than the entity they constitute.

This is the case with material structures from the macroscopic scale down to the atomic scale, where we find the greatest simplicity—all atoms, making up all molecules and ultimately all biological and inorganic objects, are made up almost exclusively of three types of particles: protons, neutrons and electrons. When we try to break apart those particles by smashing them into each other, however, we find a very rich and complex world of subatomic particles, which has ultimately been organized neatly in a single conceptual framework called the standard model of particle physics. The mathematical formulation of this model, though, is not that simple, and it seems that the upcoming experiments with higher energies may lead to revisions of the model. All this begs the question: is nature really that complicated? Can the smallest entities that make up everything really be so complex and rich in structure, with an ever-growing complexity the deeper we probe?

And that is not all! Even if the most fundamental physical theories manage to successfully explain all physical phenomena, still there will be a major mystery remaining untouched, one that has never been addressed by mainstream physics, namely consciousness. The existence of conscious experience is an undeniable fact according to philosophers, but it has remained largely unnoticed by most physicists, who seem to assume that the brain and the faculty of consciousness it supports are just a special arrangement of particles obeying the laws of physics, and finding the most fundamental physical law will entail also understanding this special arrangement. This assumption, however, is far from obvious for most philosophers.

In the remainder of this chapter we will explore a new, ontologically fundamental idea, namely, that particles, and consequently all matter in general, can be imagined as *representations* whose interactions effectively form a type of state space with a phenomenology closely resembling that of matter in physical space, with all the unusual quantum phenomena and relativity principles falling out naturally from this single postulate—that particles are representations in nature and that this is all we need to assume about them in order to derive the picture of the physical reality as revealed by physics.

The representational nature of matter thus becomes a central organizing principle of a conceptual framework encompassing one's entire knowledge of the world in a single unified body, with the understanding of the simplest forms of matter at its most fundamental level, and all more complex material structures and their interactions at successive higher-order levels. This way of picturing knowledge organized in a single system is not new; it closely matches the traditional structuring of scientific disciplines, with the natural sciences at the more fundamental level and the social sciences and humanities at more abstract, higher-order levels. The hope of philosophers envisioning science in this way is that eventually all scientific disciplines would merge into one whole, with the knowledge in the more abstract disciplines being reducible to the knowledge in the more fundamental disciplines, and ultimately to movements of elementary particles described by a single equation.

It is unclear, however, where consciousness fits in this picture, and whether it is reducible to movements of particles or it is some kind of substance entirely different from conventional matter. Most people in natural sciences are proponents of the first proposition, while philosophers and the general public are more divided and a large number of them adopt some kind of a dualist stance, i.e., they do not believe that consciousness is reducible to movements of particles.

So, where does the idea of particles being representations in their nature fit in this spectrum of views on consciousness? Since it does not postulate the existence of a new type of substance besides ordinary matter, it can be viewed as a materialist position, but on the other hand it requires a re-thinking of the traditional concept of matter, making it more akin to concepts like mind, knowledge and information, so, in another sense, it merges the two separate concepts of matter and consciousness into a single one—that of representation—which retains the properties of both original concepts. Thus, it can be viewed also as a type of identity theory, postulating identity between matter and mind, and necessitating a conceptual change with regards to the original concepts.

In order to provide some reasons as to why we should regard particles as representations, we will examine some experimental observations which are hard to explain in terms of our classical intuitions of particles as tiny balls moving in empty space. There are a number of phenomena that manifest the quantum nature of particles, and probably the simplest and easiest to explain one in terms of classical physical concepts is the diffraction of light and particles, or in some cases entire molecules, when going through tiny openings in a solid barrier. The simplest version of this setup is dubbed the double-slit experiment.

The Double-Slit Experiment

Centuries ago, when physicists started understanding the behavior of light, they asked themselves the question of whether light represents a stream of tiny particles or a packet of tiny waves propagating in empty space. Some experiments seemed to support the first proposition, while others the second one. Ultimately, physicists settled on the viewpoint that light, and electromagnetic radiation in general, has a dual nature—it comes in units, called photons, but they exhibit behavior typical of classical waves; they diffract, interfere, expand with distance travelled, etc. The double-slit experiment, where light passes through two openings and forms a striped pattern on a screen behind the openings, closely matched the physical pattern of waves on the surface of water passing through two openings and forming a pattern in the space behind the openings, and was considered a strong argument in favor of the wave nature of light. The question remained regarding the medium in which those waves propagated, by analogy with water, whose surface mediated the propagation of the water waves. Physicists postulated the existence of such medium for light, and called it the 'ether', but even the most precise measurements couldn't reveal any effects due to this medium. The speed of light was always found to be the same no matter how the experiments were done. These findings proved to be the end of the idea of an ether and eventually sprouted the theory of relativity, postulating that light travels at a constant speed with respect to all reference frames, i.e., no matter what is the speed of the observer.

So, light turned out to have some very unusual and counterintuitive properties. Some of them are revealed by a setup like the one in the double-slit experiment. More precisely, in this setup there is a source of coherent light, made up of photons of the same wavelength, like the light produced by a laser. The source is limited to emitting the stream of photons in one direction (although if it is not, we can simply ignore the photons that go out in the other directions that are of no interest to us). At some distance in the direction of propagation of the light ray there is a barrier which is opaque to light, except for two tiny vertical openings (slits). These openings need to be sufficiently narrow, on the order of about the wavelength of the incident light, and both need to be in the path of the light beam. When these conditions are met, if we put a screen (any kind of reflective surface) behind the two openings, we will see a pattern of stripes forming from the light passing through the slits and reflecting off the screen. This is basically the setup in the double-slit experiment.

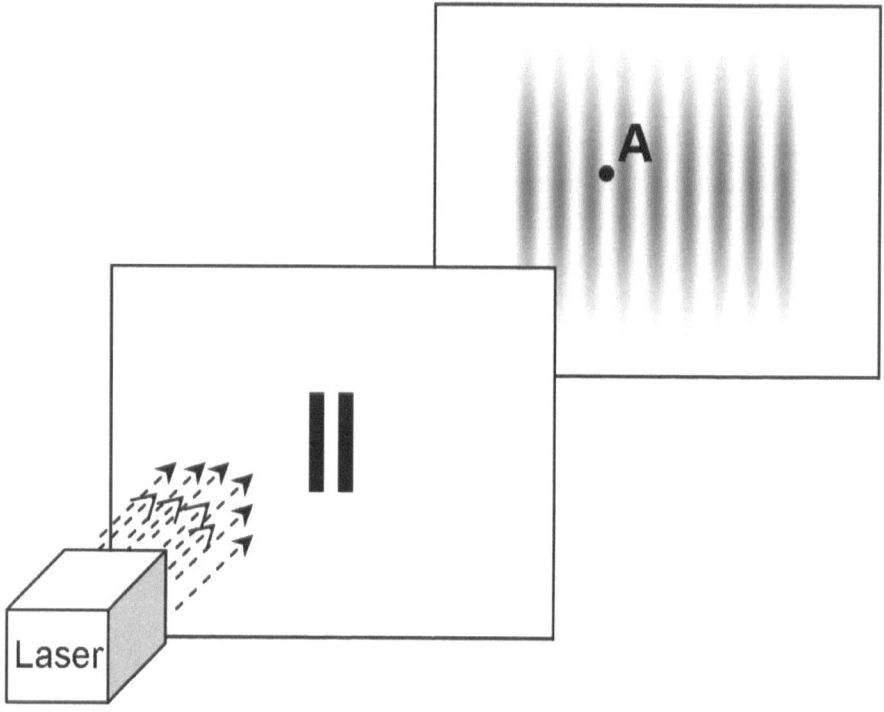

Figure 1a: The setup of the double-slit experiment and the resulting striped pattern when both slits are open. Point A on the reflective screen is not illuminated by light, i.e., no photons arrive there.

Now, the remarkable fact is that this pattern of alternating dark and bright stripes (with smooth transitions between them) is not a simple sum of the patterns formed when only one of the two slits is open and the other one is covered. In those cases we have a single wide stripe of light right behind the open slit.

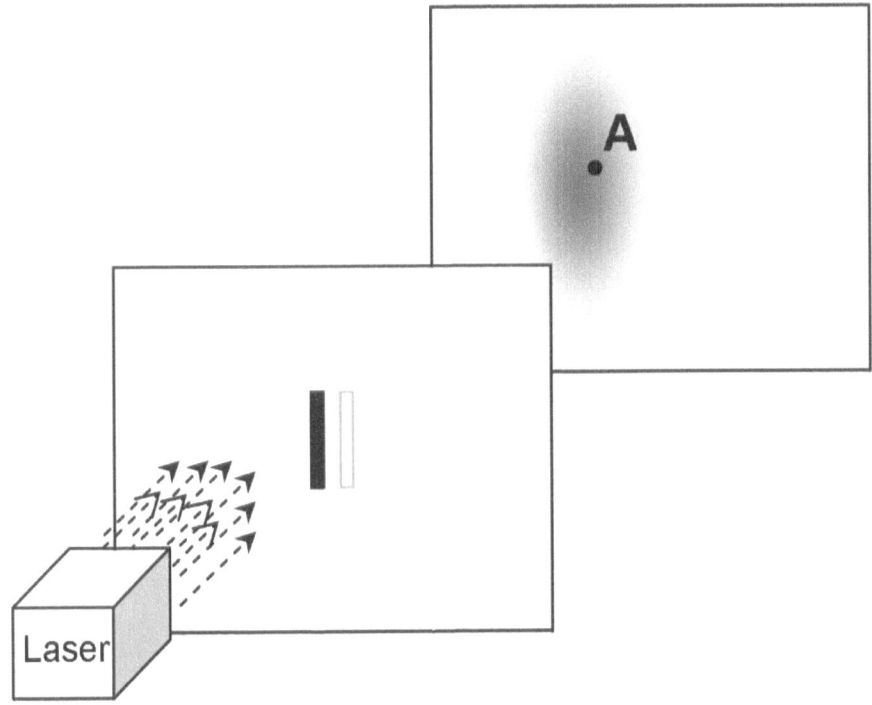

Figure 1b: The setup of the double-slit experiment when only one slit is open. Point A on the reflective screen is illuminated by light in this case.

When both are open, we do not get two wide stripes, one behind each slit, but a multitude of narrow stripes covering the space behind the two slits. This is the classical signature of interference of two waves originating at the positions of the two slits and propagating towards the screen behind them.

This picture can still be explained in classical terms, by assuming that some photons squeeze in through the first slit and others through the second slit, and they interact with (bounce off) each other on the way to the screen, forming the observed pattern. This can explain also the odd fact that there are some locations on the screen (like point A in the figure) which reflect light when only one slit is open, but are dark when both slits are open! What this means is that opening the second slit somehow expels the photons which arrive at this location when only the first slit is open.

This was the generally accepted explanation until physicists learned how to make the stream of photons weaker and weaker, down to the point when photons were emitted from the source one by one. If that is the case, surely there is always only one photon travelling at any time from the source through the openings to the screen, and there is nothing to interfere with. But even in this case, after a large number of photons were registered at the screen, the same pattern of stripes formed as when a stream of many photons travelled simultaneously from the source to the screen!

This is quite a remarkable observation. Even when only one photon travels from the source through the slits to the screen, opening the second slit still makes the photons change their trajectories in a way that prevents them from landing in some of the spots where they were landing when only the first slit was open. How can this observation possibly be explained? In the classical conceptual framework of particles as tiny balls moving in empty space, mentioned in the previous section, there is only one way out—one needs to postulate the existence of some kind of 'unreal' photons, which are sort of 'twin' particles to the ordinary photons, and which interact only with them and only in a very specific way. In the context of the double-slit experiment, we need to assume that as an ordinary photon is emitted from the source, simultaneously one or more 'unreal' photons are emitted too, and some of them pass through the same slit through which the real photon goes through, but other pass through the other slit, and on the way to the screen these other photons interact with the real photon and change its trajectory. What this means, in general, is that for every event of emitting or absorbing a real particle there is a large number of events of emitting and absorbing 'unreal' particles of the same type, effectively forming a multitude of universes made up of unreal particles, which in some way exist 'in parallel' to the real one and interact with it only in some special ways. This is the so-called 'many world' or 'parallel universes' interpretation of the phenomena described by quantum mechanics.

This interpretation handles elegantly some awkward features of quantum mechanics, like the collapse of the wave function and the wave-like diffraction phenomena described above, but this comes at the expense of assuming the reality of a vast number of universes existing besides our own universe. In fact, since every interaction between two particles is describable only in probabilistic terms, each such event would require a branching off in the course of events making up the universe, so each interaction effectively generates a number of entire new universes, one for each possible outcome of the interaction. Since such interactions happen by the zillions every second, this means that an unfathomable number of new universes are created at any instant in time—quite a difficult proposition to swallow for common sense!

So, is there some other way to explain the awkward findings from the double-slit experiment? How could the photon (or electron, or a whole fullerene molecule) possibly 'know' whether one or two slits are open and behave accordingly? In the framework of the classical picture of matter this seems impossible, so maybe we need to reconsider our basic assumptions regarding particles and space?

Indeed, the other way out is to assume that the word 'know' in the previous question is not a mere metaphor, but describes the nature of the particles in some more immediate sense. In other words, the photon must indeed 'know', in some sense, how many slits are open, or more generally, the configuration of particles in its environment. Stated in a more technical language, this means that the particle must contain a representation of its environment. This is the key insight which we will discuss in the sections that follow, and which brings about a completely new way of thinking about matter and space, and consequently about life and the associated with it phenomena of mind and consciousness, and ultimately about the totality of all things that exist, what we would call our world, or our universe. After the reformulation of some of the basic quantum phenomena in terms of this new idea we will return to the double-slit experiment and we will try to cast the explanation of what is going on in terms of our new understanding of matter as representations.

Although we come to think of particles as representations in order to resolve the puzzle introduced by the behavior of the photon in the double-slit experiment, we will see that this assumption has the power to resolve not just this particular conundrum, but many others (including consciousness), which makes it not just a patch on one particular aspect of a scientific theory, but a truly fundamental paradigm shift in our entire conceptual system which we use to understand everything that happens around us, both in terms of scientific knowledge and of general layman conceptions of the world. So, let us try to imagine in a greater detail what does it mean to think of particles of matter as representations.

Particles as Representations

A representation can be defined along the line of something that resembles something else but is physically distinct from it. In other words, a representation of something has a relation of similarity to it at an abstract level, but a relation of difference to it at the physical level. For example, a picture of a house is a representation of the physical object 'house'. The picture has the property of reflecting a very similar pattern of light as the house when viewed from a certain vantage point—this is the relation of similarity between them, but the picture is a physically different object than the house, i.e., both are made up of different sets of physical particles. In a similar fashion, we can assume that the arrangements of particles in the human brain form patterns which in some way are representations of external objects, although, unlike the case with the picture we do not know exactly how the pattern corresponds to the object. Still, we can surmise that the particles making up the brain form representations of the macroscopic environment because our conscious experiences formed inside the brain are nearly always *about*, or representing, objects in the environment. So, in very general terms, we can say that the idea of representation is based on the notions of similarity and difference: they both need to be present in a representation.

Conceptually, if we want to imagine particles in terms of representations through the notions of similarity and difference this would represent a fundamental change in our conceptual framework and we would need to be prepared to abandon also some of the other classical notions forming the familiar picture of physical reality, like space, movement in space, mass, etc. Thinking of particles as representations has consequences also for all these other notions, and they would need to be reconsidered and understood within the framework of interacting representations rather than the classical framework of the physical reality based on everyday experience. Thus, space might not be the empty container, or stage, where particles (as representations) reside, but it may be an epiphenomenon arising from the rules according to which particles interact. Movement in space may not simply be a displacement of one and the same particle with respect to other particles, but it may also require a transformation of the internal state of the particle, as it is the case with the trajectory of a state in an abstract state space. Mass may not be simply a tendency of a particle to resist displacement in case of an action by an external force, but it may have something to do with the nature of the representation and its internal configuration. Forces may not be some (mysterious) actions at a distance, but an epiphenomenon due to the 'awareness' of the representation of its environment, somewhat like the way moving people are aware of obstacles in their path and they circumvent them.

All these suggestions will be explored in turn in the following sections, leading to a new, mutually consistent conceptual framework describing physical reality based on the central idea of thinking about particles as representations. First, we will examine what happens in a system of interacting representations when they change their states with respect to each other. This will be our first attempt at replacing the picture of particles as tiny balls with picturing them as representations in some more abstract sense.

Physical Space and State Space

We stated that we will imagine a particle as an object whose content constitutes a representation of its environment. Let's call the particular representation of a given physical environment at a given moment in time the state of the particle. Also, we will assume that the particle can retain its identity while going through different states, i.e., it does not transform into a different particle when some of its properties, like position or energy, change. This is the natural assumption in physics, where a particle retains its identity as long as its fundamental properties, like charge, spin umber, lepton number, etc., related to laws of conservation, remain unchanged.

Being a representation of its environment, a change in the particle must correspond to a change in its environment, which can happen logically only in two possible ways—either the representational configuration of the other particles in the environment should change while the particle remains stationary, or the relative position of the particle with respect to the particles in the environment should change (or both, which is typically the case). The second proposition is true because any displacement of the particle with respect to other particles necessarily entails a change of the environment of the particle and therefore a change in its internal state, which, as we assumed, constitutes a representation of that environment.

Let's focus on this proposition. If any movement of the particle is associated with a change in its internal state, then we can associate an internal representational state of the particle with each location in space, just like this is done in any abstract state space, like in the pressure-volume graph of a gas, for example.

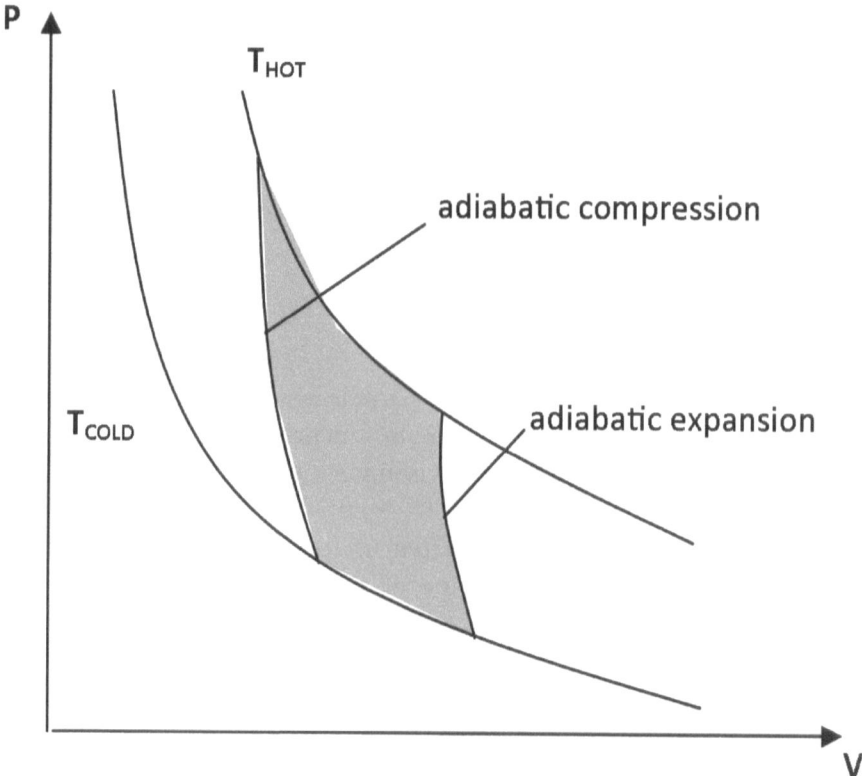

Figure 2: The pressure-volume graph of a gas with a Carnot cycle—an example of a two-dimensional abstract state space.

Thus, the single assumption of a particle's state being a representation of its environment necessarily endows space with the character of a state space, where each point in space is associated with a different internal state of any particle that may occupy that location.

This is quite a remarkable conclusion! It means that space is not a pre-existing stage where particles reside, independent of their presence inside it, but it owes its existence to the particles and their internal states. We can even imagine the particles as existing in some abstract sense not contained in space, just existing somehow independently of each other, and the relationships between them, due to the fact that they are representations of each other, producing an appearance of distances between them and effectively a space which they populate. It should be noted that the space itself does not remain static, as the one in the pressure-volume example, but changes constantly with the transformations of the particles. This, however, does not invalidate our conclusion, since basic properties, like distances, may remain invariant under those transformations.

The nature of space as a type of state space arises from the assumption of particles being representations *of each other*, not just representations of something else, i.e., the relationship goes both ways: if a particle contains a representation of the state of another particle, then the opposite is true—the other particle contains a representation of the first one. Thus, the representations of the different particles are in some sense linked to each other and they change in synchrony, in an orchestrated, non-arbitrary fashion.

This regularity of the changes in their internal states produces the stability of some aspects of the relationships between them, which gives rise to permanence of the transformations of these relationships, e.g., how much time it takes to transform from one state to another, and this in turn produces the epiphenomenon of particles moving in space and interacting with each other according to fixed laws. The interactions usually occur when the particles get in spatial proximity of each other, i.e., when they collide, but as we know from some physics experiments, this is not necessarily always the case. In some cases it is possible for two distant particles to have a shared state, in which case they act as one entity called by physicists an entangled state[1].

So, we need to imagine space as some kind of state space, at least locally, at short distances. We should note that distance in state space is correlated with difference in the internal states of two particles—the higher the separation between two particles in state space, the bigger the difference between their internal states. Now, if we assume that representations transform always in one and the same fashion, i.e., they transform in a deterministic way, then it follows that the greater the

---

[1] This is what Einstein called "spooky action at a distance".

difference between two representations, the longer it takes for them to get reconciled, i.e., to converge to a single shared state.

Let us try to picture this process in terms of particles in state space. When two particles have widely different states, they must be at some distance in state space. When they engage in an interaction which eventually brings their two states to a single shared state, this means in terms of their locations in state space that they start moving closer to each other and eventually meet at a single location at the moment when they reach the shared state. This picture already looks very much like two particles experiencing an attractive force between them and colliding in space. So, the simple assumption of particles being representations of each other leads to the picture of space as state space and in turn of interactions between the representations producing the epiphenomenal appearance of particles moving and colliding in this space, subject to forces of attraction and possibly repulsion.

We will consider the epiphenomenal appearance of forces in greater detail later on, so let us focus now on the regularity in the transformation of the representations. If we assume that the transformations of the internal states of the representations occur in some sense regularly, i.e., a pair of interacting representations in exactly the same states on two different occasions will transform in exactly the same way, then it follows that the movements of the particles in space would be regular in some ways. More specifically, if we accept the proposition that all elementary particles are essentially the same, which seems to be widely accepted by physicists, that means that the way they move would be determined entirely by the distance between them, i.e., they will be subject to highly regular laws of movement. However, since the particles contain a representation of their environment, which is different from another particle's environment in the general case, they must differ in some ways, which would add some uncertainty to their movements in space. Thus, the similarity component of the representations is responsible for the permanence of the laws of physics, while the difference component produces deviations from the average values, conceptualized as the fundamental and unavoidable fluctuations of the particles' states.

So, the fact that the particles (i.e., representations) are pretty much the same, except for the part related to the specifics of their environment, means that the movements of the particles in state space are highly predictable and can be described by laws. What's more, since the transformations occur with the same speed every time, this produces the appearance of the particles moving with the same speed through state space. This must be valid at any time for particles whose representations are completely dependent on their environment, i.e., they reflect only other representations in the environment and nothing else.

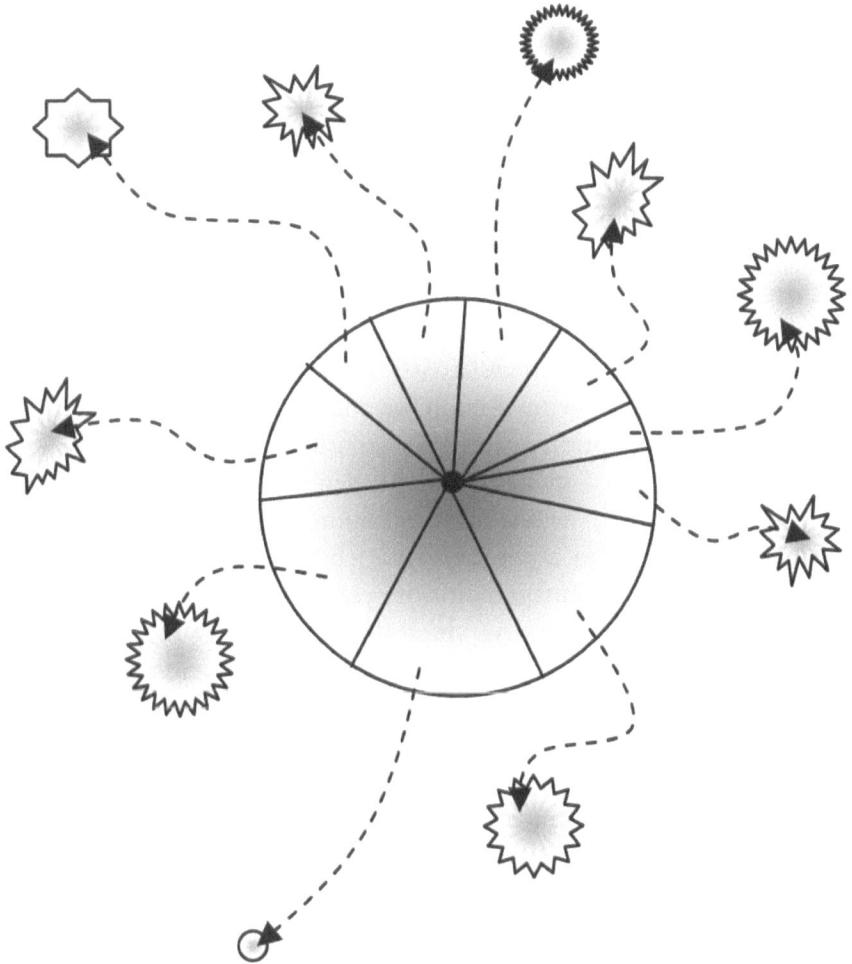

Figure 3: Particle as a composite representation (the big circle in the middle) consisting entirely of representations of other particles in the environment (the oddly shaped blobs surrounding it). Arrows represent the outward-directed representational relations. In this case the particle is moving at the speed of light.

As we will see in a moment, representations can be self-reflecting, i.e., directed towards another part of the same representation instead of towards another representation in the environment. In this case it is possible for them to transform with different speeds, and consequently for the particles to move with different (lower) speeds in state space. In the absence of such self-representation, however, the transformation of the particle is fully determined by its environment, and since a) distance in space reflects difference in the representational states and b) transformations are deterministic, it must transform always with the same speed, which looks like movement of the particle with a constant speed through space. So we arrive at the well-familiar picture of the constant speed of light, which is the mandatory speed of movement for massless particles! The assumption of regularity of the representational transformation together with the assumption of state as state space in this way entail the constant speed of movement of massless particles.

Next, we will consider the relativistic effects due to the constant speed of light and in the following section we will attempt to explain them in the light of our assumption of the representational nature of matter.

The constant speed of light, independent of the reference frame in which it is measured, is one of the counterintuitive facts discovered by physicists more than a century ago, which is hard to visualize and comprehend from the vantage point of everyday experience. It lies at the heart of the theory of relativity, which describes among other things the gravitational interactions between material objects and endows space with the property of having a shape, i.e., it can be stretched and compressed, affecting the straight line trajectories of objects moving through it, even in the case of light. Thus, light bends when passing close to massive objects, like large stars, instead of moving along a what would be geometrically a straight line if there was no star in the vicinity.

Even more unusual are the consequences of the constant speed of light in all reference frames for objects travelling at high velocities, close to the speed of light. The higher the velocity of the object, the larger its mass becomes according to the theory of relativity, and the consequence of this physical law is that objects cannot travel faster than the speed of light because their mass would become infinite at this speed, which is physically impossible. Another strange consequence is that the passage of time slows down the faster an object moves. Thus, an astronaut travelling at a speed close to the speed of light would age much more slowly than a person staying stationary with respect to the original reference frame at the time of the departure of the astronaut, leading to the so-called paradox of the twins. If one of two twins takes off on a journey through space with a speed close to the speed of light, he or she can return back to Earth at a time point much later than the departure time, e.g., a thousand years later, when the other twin would be long ago dead. This is indeed possible and verified by experiments with atomic clocks, although such a journey is beyond the capabilities of present-day technology.

The third unusual consequence of the constant speed of light is the length contraction in the direction of the movement of an object. At slow speeds this contraction is insignificant, but at speeds close to the speed of light the objects, and space itself, indeed 'shrink' along the direction of movement. This phenomenon cannot be detected in the reference frame of the moving object, since everything in this reference frame shrinks, including any ruler or other instrument for measuring lengths, but it can be observed from a reference frame which has remained stationary with respect to the moving reference frame.

So, we see that the physical phenomenon of the constant speed of light in some sense redefines space, making it appear more like an entity dependent on the behavior of the objects inside it, rather than an immutable stage where the objects are situated and behave independently of the space around them, which is the intuition due to common sense. Still, physicists do not have a definitive explanation as to *why* the speed of light is constant. This is a generally accepted fact based on empirical observations, but it cannot be derived from some more fundamental postulates. To do this, an all-encompassing theory of everything would be required, which reconciles relativity with the other theories describing physical interactions, namely, quantum theories of the three other forces besides gravity—the electromagnetic, the strong and the weak forces. We will deal with the interpretation of those forces in terms of representations later on, but now we will try to understand what are mass and the force of gravity in terms of our picture of particles as representations situated in some type of state space. In the following section we will examine how the three phenomena associated with the constant speed of light that we just described arise in the framework of particles as representations.

What is Mass

We stated already that the representations can be referring to completely different representations or towards parts of themselves. Let us think about what are the consequences of a representation being directed towards itself. This means that one part of the representation is reflecting another part of the representation and vice versa, since the relationship is bidirectional. Thus, the mutually reflecting parts of the representation constitute a unit which a) is not reflected in external representations and b) has a tendency to remain stable, due to the fact that the mutual representation is self sufficient and would need external influence in order to transform in some ways. On the other hand, this is only a part of the whole representation constituting a single particle, so there are other parts which reflect external representations in the environment, but which also need to be integrally connected with the self-reflecting part of the representation (we assume that in order for the representation to persist as a unit, it needs to maintain integrity by smooth, gradual transformation of the representation, which would enable it to stay one whole when subjected to external influences).

So, what would be the effect of the presence of a self-representation as a part of the whole representation making up a particle? Since it has a tendency to remain stable and it is connected to the other parts of the representation, it would have the effect of slowing down their transformations due to changes in the environment. This could be described also as some sort of a 'drag' in the speed of transformation of the representation as a whole. In terms of movement in state space, this would mean movement with a slower speed than the maximal (or normal) one, i.e., self-representation acts exactly as mass, slowing down the movement of particles.

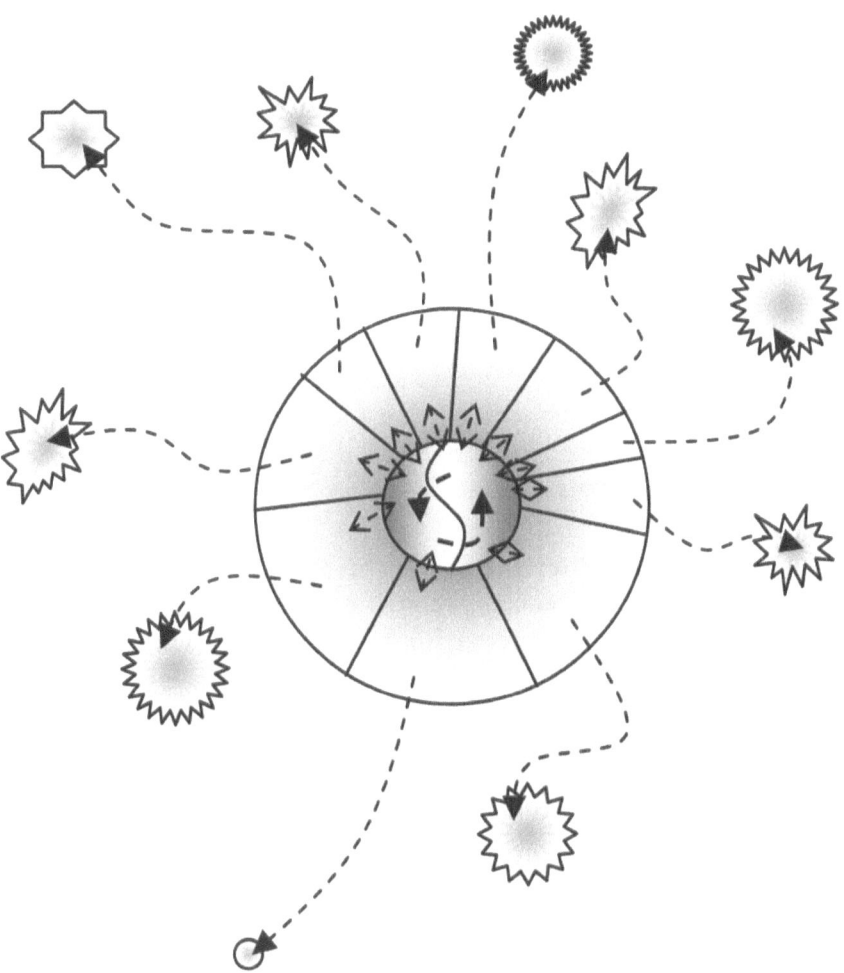

Figure 4: Particle containing a self-representation (the small circle at the center of the big circle). There are also representational relations between the self-representational core and the outward-directed representational components in the periphery of the particle (signified by the bidirectional small arrows). In this case the particle is moving at a speed lower than the speed of light due to the drag of the self-representational component.

In the extreme case of complete self-representation the movement would cease completely, but this seems to be an impossible proposition since the particle would have to go from a state of reflecting its environment to a state of complete internal reflection, and the only way this can happen is under the influence of external forces. However, if it is subject to the influence of external forces this means that it continues to reflect its environment, so it cannot 'close' itself completely. This reasoning is very much in line with the physical impossibility of attaining the absolute zero temperature—the state of complete cessation of any movement, and therefore, any transformation. Physically, it is possible to get arbitrarily close to it, but never to attain it completely.

So, we arrive at the conclusion that the degree to which the particle representation is self-reflecting determines the 'drag' on its transformation, and effectively its mass. If a higher proportion is self-reflecting, this means higher mass, and vice versa.

In terms of the movement in state space, if a representation is not subject to external influences, the two tendencies—of transformation due to the part reflecting the environment, and of stability due to the self-reflecting part—would settle in some state of balance between them, which would result in a constant, but lower than the maximal, rate of transformation and consequently in a steady movement through state space. This is the familiar inertial movement of physical bodies when they are not subjected to external forces. If the representations in the environment change in some irregular way, which is not anticipated by the gradual transformation of the particle's representation, then it would need to reflect those changes and therefore its representation would need to change its natural course of transformation, i.e., it would experience external forces.

Now, consider what happens when the environment changes the course of the transformation of the representation. The effect of external influences, or forces, is to make the representation less internally reconciled, i.e., to try to disintegrate it to a greater degree. If the self-reflecting part of the representation is to remain intact, it needs to resist this tendency, i.e., it needs to increase its degree of integration and couple the two parts stronger. Thus, the portion of the representation which reflects the environment would become more irregular and less reconciled, while the portion that constitutes the self-representation will become more strongly coupled. So, effectively, the balance will be tipped towards the self-representation and it will become more dominant than the external representation, making the mass of the particle grow larger. The less reconciled part of the representation would transform more rapidly, which would have the effect of a faster movement through space. Thus, the effect of exerting a force on the particle is both faster movement and increase of its mass, in line with the principle of relativity described in the previous section (of course, it is possible to exert a decelerating force on the particle, in which case the reasoning goes the opposite way).

We saw how mass can increase with increasing speed of movement of the particle. Now we will consider the effect of length contraction in the direction of movement of the particle. To do this, we need to remind ourselves of our picture of space as state space engendered by the interactional relationships between the particles. We stated that distance in space is proportional to difference in the internal states of any two particles. Let us consider now what happens when two particles move with respect to each other when they are subject to external forces, as in the cases just described.

If a particle moves towards another particle, subject to an external force produced by this particle (and possibly others), this means that its representation of the environment is influenced by the representational state of the other particle, i.e., it is in an interaction with the other particle. Now, when two different representational states are engaged in an interactional relation, they must necessarily be representations of each other, as we stated before, but when their representations interact in such a way that the distance in space between them gets shorter, this means that their representations must be getting more similar. This transformation towards greater similarity represents the converging movement in space, but in order for them to become more similar, there must be a component in the representations which drives this transformation, i.e., this component must in some sense anticipate the greater similarity; it must be a similarity itself with an even greater degree of reconciliation.

This is how the movement of two particles towards each other is different from movement in parallel—the representations have an anticipatory component which is going to make them more similar in the future. Because of that, they have an additional similarity potential at present than if they just transform in a coordinated way (in the case of parallel movement). The anticipatory component would affect the metrics of the epiphenomenal state space associated with the particles in a way that would make the length of the space intervals in the direction of movement shorter compared to intervals perpendicular to the direction of movement (these intervals will separate particles moving in parallel) due to the extra similarity between their states.

There is one more additional consideration in this line of reasoning. Length is a property of a solid object situated in space, which constitutes a system of multiple particles, not a single one, and when all those particles have the same anticipatory component due to their collective movement in space, the properties of the space in the direction of their movement are changed. Length is measured by sending a light signal along the measured direction and receiving its reflection after some period of time, but due to the presence of the anticipatory component in every particle in the inertial frame of the moving body the properties of the light signal are changed too and the interaction between two distant particles that they mediate occurs a bit quicker from the point of view of an inertial frame at rest, producing the illusions of shrunk space (and therefore shorter length) and time dilation in the moving body frame.

In summary, we see that the peculiarities of the transformations of representations influencing each other affect their properties and in effect the space that separates them. Since any transformation of a representation must be causally induced by another representation and the direction of change for the transformation is not arbitrary, the direction of change is linked to the degree of similarity between any two interacting representations and therefore affects the distance relations between them. More specifically, space shrinks when two representations are in the process of becoming more similar.

Let us try to extend this principle to a system of a very large number of particles engaged in an interactional relationship with each other. We stated that the part of the representation of the particle which reflects the environment has an anticipatory component which affects the distance relationships between particles moving in space. Earlier, we also posited that the two parts of a particle's representation—the one reflecting the environment and the self-representation—are tightly connected, since they constitute a single unit which we call a particle. So, the two parts must influence each other. If one of them is subject to change, the other one would experience change too, although to a lesser degree. This means that when the part representing the environment is subjected to external forces in a way that makes it more similar to another representation in the environment (which, as we said, is expressed as a movement towards the other particle), then the self-referencing part also gets modified a little bit in the same direction, i.e., it becomes more similar to the self-referencing part of the other representation. This would be experienced as a slight force of attraction between the two particles—what we call gravity!

To state it another way, the interactions between particles affect their self-reflecting states, responsible for their mass, in a way to make them more similar to each other. This effect in terms of state space has the result of pushing the particles closer to each other, i.e., it acts as an attractive force, very much like gravity. So, gravity emanates basically from the shared history of the particles' states due to their interactions. In a sense, it is like a collective memory of their past states to which they tend to return with a slightly higher likelihood than to go to completely novel states.

In terms of the metrics of the space populated by a large number of particles, given that they move chaotically the proportion of them moving towards each other must be about the same as the proportion moving away from each other. Thus, the net relativistic effect of the movements of the particles is zero, since on average their states do not get more similar or more different from each other. However, due to the memory effect which manifests itself as the attractive force of gravity, there is a tendency for the self-reflecting parts of the representations to get more similar, and this affects the metrics of the space near the large clump of particles. If another particle comes in the vicinity of this clump, the other representations in its environment (which belong to particles from the clump) would share a greater degree of similarity and their states would collectively influence the state of the new particle with a stronger force than what would be the case if those states were very dissimilar (i.e., not sharing a common history and therefore not experiencing the attractive force of gravity). This larger impetus to change the state of the new particle acts exactly like the anticipatory component in the case of two particles moving towards each other, since its effect is to bring the particles' states closer to each other. Correspondingly, it affects the metrics of the space in the vicinity of the large clump of particles, making it seem 'shrunk' in the direction of the clump. So, we arrive at the familiar to the physicists picture of the curved space due to the presence of mass.

We imagined already on many occasions how representations influence each other even when they are at some distance in space. In fact, we stated that space is an epiphenomenon and we need to imagine representations somehow not being situated in any kind of space, but simply existing independently of each other, and the rules and regularities governing their interactions giving rise to the epiphenomenal appearance of space in which they are situated. Thus, the possibility of interactions between representations situated at some distance in space produces the concept of force acting at some distance on another object. We will discuss this concept in the following section.

What are Forces

The traditional concept of force in physics is derived from our everyday experience of the macroscopic world and not from the realm of the subatomic scale. In the macroscopic world, causal interaction between two bodies usually requires that they are in contact and one of them pushes the other one and causes it to move. In some cases, however, it is possible for the pushing to occur when the two bodies are not in a direct contact but at some distance from each other. In those cases it seems like there is some invisible connection between them which transfers the movement of one of the bodies to the other, and this phenomenon gives rise to the concept of a force acting at a distance. The most familiar invisible forces are those of electrostatic repulsion and attraction between two electrically charged objects and the magnetic repulsion and attraction between magnetized objects.

It is obvious in those cases that, somehow, one object is able to repel or attract the other object when it is at some distance, so something must be emanating from it and arriving at the other object. Historically, this idea persisted also in the studies of ever smaller and smaller scales, down to the realm of subatomic particles, and it culminated in particle physics in the idea of particles-carriers of forces. So, according to modern day particle physics, forces are due to the exchange of one type of particles, called bosons, between the other type of particles, called fermions. The fermions are the particles making up all macroscopic objects, and all of them have mass and generally speaking cannot occupy the same location in space, while the bosons have the unusual property of being able to occupy the same location in space without bumping into each other. Also, some of the bosons have mass, but others, like the photon, don't.

There are four fundamental forces, i.e., four different ways in which particles can interact with each other, at relatively low energy scales, and it is believed that at sufficiently high energy scales all those forces become a single, unified force. There are a number of theories proposed on how to unify three of the forces (the electromagnetic, the weak and the strong ones), which are somewhat controversial, but the force of gravity is the odd one out in modern physics theories, and the theories describing its unification with the rest (quantum gravity, string theories, etc.) are highly controversial.

So, is the picture of massless (in the cases of the electromagnetic and the strong force) particles exchanged between the fermions really what happens physically at the subatomic scale? Or is there another way to think about forces? This question is motivated by our earlier picture of what happens in the double-slit experiment. There we saw that instead of thinking of photons passing through slits we can think of the particles as if they are possessing some kind of 'knowledge', or a representation, of the environment, which makes them act in a coordinated fashion. This idea renders the particles-carriers of forces the status of epiphenomena: the mass-possessing particles engaged in interactions behave in a coordinated way which produces the appearance of something being exchanged between them, but in reality nothing physical (such as a photon or a gluon) may be moving from one particle to the other. Instead, the coordinated changes in the fermion particles could be seen as the exchange of something that we may call information, but which is not physical in the sense that the fermions are.

Let us consider this idea in the light of our picture of representations situated in state space. The fermions are mass-possessing particles, so, according to our picture, their representational state has two components—one is self-reflecting and the other one is reflecting other representations in the environment. Those two components are tightly coupled, meaning that any change in one of them would influence the other one and exert pressure for change there. We saw how this leads to a number of phenomena described by the theory of relativity. To see how the changes in two particles containing representational components of each other become coordinated we need to focus on the fact that mutual representational relations may be reconciled to different degrees, i.e., representations may possess different ratios of the degree of similarity vs. the degree of difference in the relationship.

So, let us imagine two particles that represent each other, and let us think of the representation as being originally in a well reconciled state. This means that the representation (or the configuration of the internal state) of the first particle matches well the representation (configuration) in the second particle, in the component responsible for the mutual reflection of their states. However, there are other components which relate to other particles in the environment, and they are in a constant transformation since the particles always move and their configurations in space change incessantly. These transformations would influence the state of the mutual representation component, since all the representational components of one particle are tightly coupled. What's more, the influences would be different, since the transformations in each particle's environment do not match perfectly (in the general case). So, there would be pressures for change on the mutual representation component on both sides (in both particles), but these pressures would differ qualitatively on each side, so they would tend to disrupt the reconciled state and to make it less reconciled.

On the other hand, the mutual representation would tend to stay reconciled, since it is a mutual relationship, so it would either try to return to the old configuration or try to find a new well reconciled configuration. Eventually, such a configuration may be found and the representation may become well reconciled again. So, we see that the representations always move from a state of being well-reconciled through a state of being less reconciled and back to being more reconciled again. The ratio between the degrees of difference and similarity always fluctuates somewhere between the two extremes (and if it reaches one of the extremes, it means that either the two particles merge into one—total similarity, or they disengage and do not interact any more—total dissimilarity). How do these constant fluctuations affect the states of the particles and effectively their positions in state space?

The constant oscillations in the representational states would produce constant oscillations in the positions of the particles in space—the familiar omnipresent 'jitter' that is observed in the microscopic world of the particles. Also, the transition from one well reconciled state to another through a less reconciled phase would produce a coordinated change in the states of the two representations. It would be coordinated because at the start the behavior of the two particles it is coordinated (due to the mutual representation in a well reconciled state), then it goes through an uncoordinated, more erratic phase, but the end state is again coordinated behavior. So, it would look as if the particles affect and change each other's behavior in the time period of the uncoordinated phase. In terms of movement in state space, this change in behavior would look like a change in the direction of movement. For example, if the particles' states were transforming initially in a way that would make them more similar, this would represent movement towards each other in state space, but the erratic phase would disrupt this coordinated transformation and would make it a different one. This would look like the two particles recoiling and continuing on different trajectories in state space. This picture is very much like the one of two particles exchanging a photon as depicted in the Feynman diagrams!

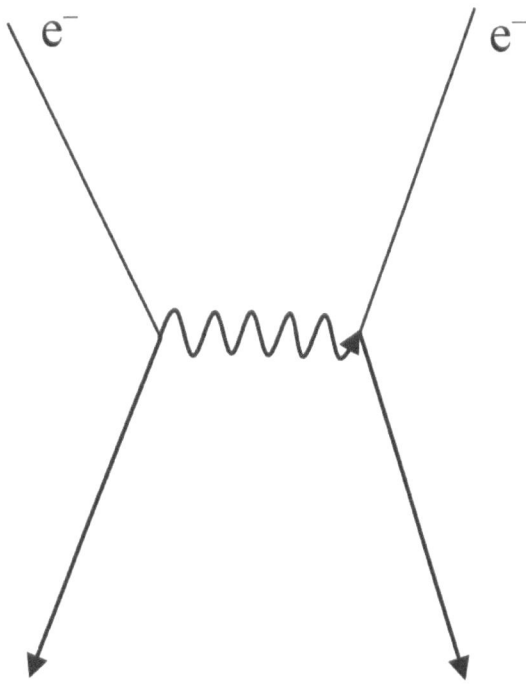

Figure 5: Feynman diagram of the simple event of an exchange of a photon between two electrons.

So, we conclude that disruptions to the mutually representative state produce coordinated changes (snaps to a new configuration) which look like exchanges of particles of force. Can we say something about the different ways this process can occur? We stipulated already that due to pressure from the other parts of the particle's representation, the mutually representative part may switch to a new reconciled state. This process seems to occur easily and frequently in a system of many particles, and is equivalent to exchange of a photon—the carrier of the electromagnetic force.

before the exchange of a photon   **after the exchange of a photon**

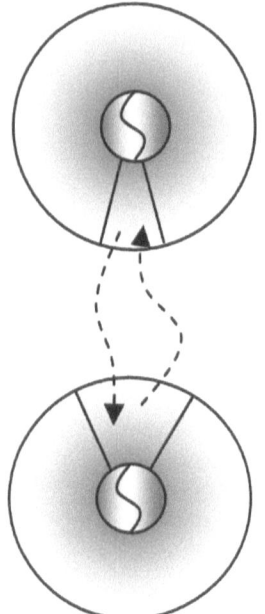 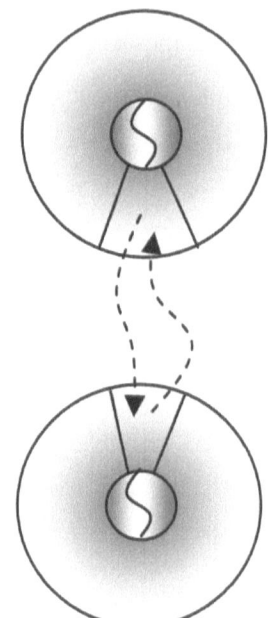

Figure 6: The mutual representational relation between two different particles before and after switching to a new reconciled state. The change in the mutual representational components is signified by the difference in the size of the segments denoting these components in the two particles. This situation depicts the action of the electromagnetic force.

The other logical possibility is for the mutually representative part to snap back to the old representational state. The effect of this would be that the particles would always be kept in the same old state (possibly with minor variations due to the pressures from the environment producing the ubiquitous 'jitter'). In terms of state space this would look as if the particles are bound to each other in a unit, and the more they are pressured to become dissimilar (and move apart in state space), the stronger they resist this pressure—and this looks just like the behavior of the quarks inside the nucleons!

So, the second logical possibility of the representation snapping back to the old state looks equivalent to the action of the strong force: it grows stronger with increasing distance and does not allow the particles to escape the vicinity of their location (i.e., to become significantly different from each other, which would endow them with their own identity). They always return to the old state and the old location.

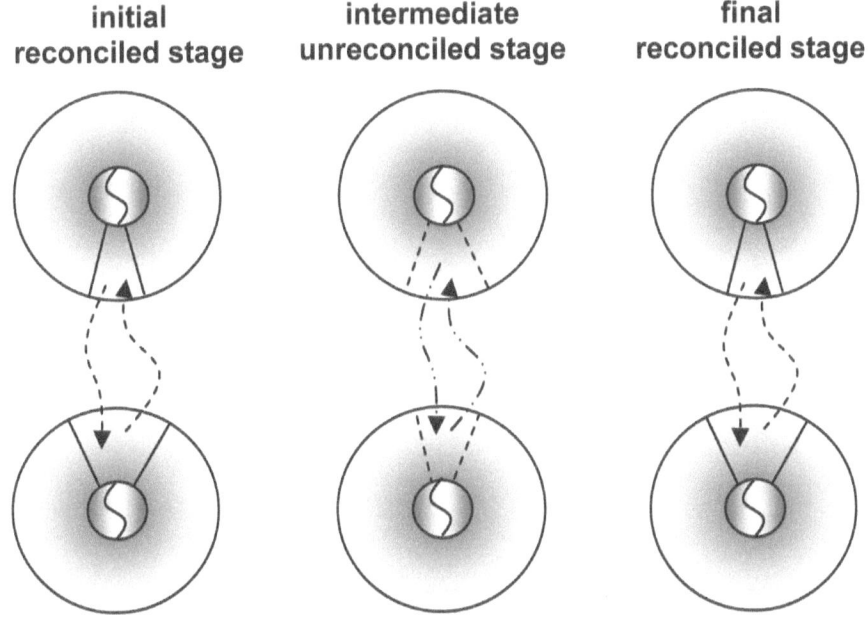

Figure 7: The mutual representational relation between two different particles going through an unreconciled state and returning back to the old reconciled state. The unreconciled phase is signified by dashed lines denoting the temporarily transformed segments. This situation depicts the action of the strong force.

Until now we considered changes in the mutual representation of two particles, but there is one more possibility which we disregarded—changes in the self-representative core of the particle; the representational component which we postulated gives mass to the particle. Although this component is stable, it oscillates too due to pressures from the other components of the global representation of the particle, and occasionally it may come close to 'snapping' to another self-representational state, much like the mutually representative state between two different particles. So, if exactly in this moment it receives the right kind of 'push' from the rest of the global representation, it may 'snap', i.e., transition to a new self-representative state. The effect of this would be a change in the identity of the particle; it would look as if the particle has transformed into another particle (or maybe has split into multiple particles)—and that's exactly the action of the weak force! So, the third logical possibility of transformation of the global representation is equivalent to the action of the weak force.

initial state

final state

(after change in the self-representative core)

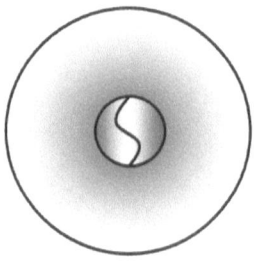
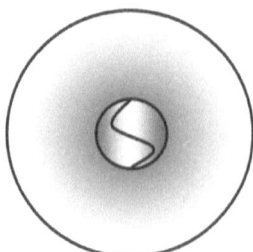

Figure 8: The self-representative core of a single particle before and after a transition to a new state. This situation depicts the action of the weak force.

Thus, we can conclude that the different possible transformations of the representational states that we assumed constitute the entities we call particles match closely the character of the three known physical forces besides gravity—the electromagnetic, the strong and the weak force. The electromagnetic force is a transition form one coordinated state to another in the representational component between two particles, the strong force is a transition back to the same coordinated state of a particle with its environment, and the weak force is a transition in the self-representative core giving the identity of the particle. Can we say also something about the ranges of those forces, i.e., how their strengths vary with distance?

Regarding the electromagnetic force, it seems that the pressures from the environment can induce a transition in the representational state relatively easy. There are many possible coordinated states and even a small push may cause the representational state to switch to another one. However, the more dissimilar two particle's states are, the less likely it is for the environment to influence them in such a way that they would make the switch. When the two particles are more dissimilar, it is more difficult for them to maintain the coordinated state, and consequently any other coordinated state, effectively making the transition to a new coordinated state more difficult. In terms of distance in state space, it means that the likelihood of making a switch to a new state, which we said is equivalent to exchanging a particle-carrier of force, diminishes with increasing distance (when the particles' states become more dissimilar). So, the likelihood of interacting through the electromagnetic force diminishes with distance, which is the same as saying that the strength of the force decreases with distance.

Unlike the electromagnetic force, the strong force, as we found, increases with distance for the reasons stated earlier. However, the likelihood of finding particles interacting through the strong force also decreases as the separation between those particles increases. The particles that always return to the original state must be only very weakly affected by their environments and mostly affected by each other, so that their states do not grow dissimilar enough to transition to a new configuration of coordinated states. So, they must always stay very similar and tightly coupled, which allows any external influences to be counteracted and cancelled completely. This means that their positions in state space must always remain very close, which is exactly the case with quarks bound up inside a nucleon. So, the range of the strong force is typically very short, and we do not find particles interacting through this force which are far apart.

With regard to the weak force, we stated that the self-representative state giving the identity of the particle needs to receive the right kind of 'push' from the environment in order to transition to a new coordinated self-representative state. This 'push', therefore, must be strong enough and also in the right direction. This is more likely to happen when the representations of the environment are more coordinated, and they are more coordinated when they are more similar. In that way, the weak force is also strongest at small distances and gets weaker with increasing distance. It acts exclusively on particles packed tightly together in a small space, like the protons and neutrons in the atomic nuclei. Unlike the other forces, however, its strength depends also on the arrangement of the particles in the environment. The more regular this arrangement is, the more likely is the action of the weak force.

This peculiarity of the weak force has the effect of imposing an upper limit on how many particles can be packed together in the small space of the atomic nucleus. The more they become, the more they affect each other's representational state, and the more regular and coordinated their states become. This, as we said, increases the likelihood of a switch in the self-representative state of any of the particles in the bunch, which effectively would break up the bunch into separate pieces. This is the phenomenon of beta decay which becomes highly likely for the heavy nuclei and which does not allow for even heavier nuclei to form, thus limiting the number of chemical elements that can exist for prolonged periods of time.

The occurrence of a transformation in the self-representative state of a particle is much less likely than a transformation in the mutually representative states between two particles or between two particles that do not switch to a new coordinated state. That is why the weak force is called 'weak': when the interaction is much less likely to happen, it looks like the force is not acting most of the time, and consequently that it is 'weak'. The electromagnetic force type of transformation is more likely to happen, and it has been measured by physicists to be about 10^{11} times stronger than the weak force, while the strong force type of transformation is the most likely and has been measured to be about 10^{13} times stronger than the weak force.

Another peculiarity of the particles-carriers of the weak force is that they have mass, while the particles mediating the electromagnetic and the strong forces don't. This fact is also readily explained by our interpretation of the forces: a transformation in the self-representative state of a particle can change the coupling strength of the self-representation, i.e., the new state would differ in how stable the self-representation is. It would look as if the exchange of a particle of the weak force between the old mass-possessing particle (the original self-representation) and the newly created mass-possessing particle(s) (the new self-representation) has produced or eliminated some mass, from which physicists conclude that this mass must belong to the particle-carrier of the weak force. On the other hand, since the electromagnetic and the strong force interactions do not affect the self-representation, they do not change the mass of the particles and therefore the photons and the gluons must be massless.

There is still another property of the transformations from one coordinated state to another which we have not discussed until now.

Much like the transformation due to the weak force changes the strength of the coupling in the self-representation, the transformations due to the electromagnetic force may change the strength of the coupling in the mutually reflective representational component between two particles. This change, however, may have two possible outcomes: either a stronger coupling than before or a weaker one. In the former case it would make the particles become more similar, or move closer to each other in state space, and in the latter case they would become more dissimilar and move apart. In effect, the force acting between the two particles would appear either attractive or repulsive in nature, which is another property of the electromagnetic force.

So, the direction of the transformation due to the electromagnetic force constitutes what is known by physicists as the charge of a particle. It can be either negative or positive, with opposite charges attracting each other and same charges repelling each other. The charge is an intrinsic property of the particles, and we can imagine it as a specific configuration of the mutual representation component in each particle. If this configuration is antisymmetric (in some sense), the representations would partially cancel out and the difference between them would become smaller, producing an attractive force. The property of antisymmetry means that the configuration in one of the particles is the exact opposite of the configuration in the other particle, which mathematically can be expressed with opposite signs (+ and —). If the configurations are symmetric, the representations would reinforce their configurations and as a result would enlarge their influence, thus increasing also the difference component that is responsible for their separation in space. This would have the effect of increasing the difference between them, producing a repulsive force. Mathematically, the symmetry can be expressed by same signs (+ and +,— and—). This notation corresponds to the familiar assignments of signs to the charges of particles.

The strong force does not affect the mutual representations between different particles, therefore it does not have a charge. However, there are 8 varieties of gluons (the particles of the strong force) and they have another property, similar to charge, called 'color charge'. The details of how it works are somewhat different, but the basic principle is the same as with charge. The weak force, on the other hand, has no limitations on how it can affect the mutually representational component between two particles. Recall that the weak force is a transformation in the self-representational component, but this transformation may be affected also by (and affect) a transformation in the representational components between particles. So, there are three logical possibilities for the action of the weak force on the external representational components: either making them more similar (acting as an attractive force), making them more dissimilar (acting as a repulsive force), or leaving them unchanged. This is exactly what physicists find with respect to the charges of the carriers of the weak force. There are 3 varieties of those particles: the W^+, W^- and Z° boson. The first one carries a positive charge, the second one a negative charge, and the third one is neutral. They correspond to the three logical possibilities of what can happen when the self-representation component undergoes a transformation.

After this lengthy discussion of forces we can imagine what constitutes a particle even more vividly. A particle is a tightly coupled structure of individual two-way representations that influence each other. A particle with mass has its own individuality due to the self-reflective component which contains information that is not represented anywhere else.

But what about the massless particles, like the photon and the gluons? From an informational perspective, they do not contain any unique information, i.e., all the information that we may think is contained in them may be found also in the corresponding representational component of the particles that are engaged in an interaction which looks like the exchange of a photon (or a gluon). So, in a sense, we can think of them either as some kind of ephemeral entities, or as not existing at all, as was suggested earlier. Instead, we can think of the interacting particles as undergoing coordinated changes which give rise to the illusion of particles of forces being exchanged between them. It is possible to think also of the particles of the weak force in that way, despite the fact that they possess mass!

The notion of force is closely related to the concept of energy. Force refers to the qualitative aspect of the influence of one physical object on another, while the idea of energy allows us to quantify this influence and to ascribe it as a property of the physical object that performs the action on another object. In the next section we will recast too the notion of energy in our already familiar framework of interacting representations in state space.

What is Energy

In physics, the concept of energy expresses the ability of one physical object to perform an action and to change the state of another object. Originally, physicists distinguished two types of energy: kinetic and potential. The kinetic energy refers to the potential for action of a body due to its movement in space, which depends on its mass and speed of movement, while potential energy is an intrinsic potential for action which exists even when the body is at rest. This could be for example energy stored in a rotational motion of the body or in the potential to move in the direction of a gravitational gradient or in the coupling between the particles making up the body (chemical or nuclear energy), etc.

We discussed already in the previous section how transformations in some parts of the global representation of a particle exert pressure on other parts and this leads to the appearance of forces. We can extend this idea further by noting why transformations of different kinds have the effect of pressure on the interacting representations. Basically, the representations always strive to be in a reconciled state. This stems from the very nature of a representation—it is a mutual relationship of similarity in the content of two distinct entities. If the representation is less reconciled, the degree of similarity in its content on both sides is lower, and consequently it is less of a representation. In the limiting case of no similarity at all, the representation disappears, i.e., it does not exist at all. Thus, every representational relationship tends to increase its degree of similarity in the absence of any external influences. This can be stated in another way as saying that the existence of something is preferred over the existence of nothing.

So, given that the tight coupling between the two sides of a representational relationship is the preferred state which all representations strive to attain, this gives us the reason for the existence of forces and energy. In a system with a large number of particles engaged in representational relationships any individual relationship strives to increase its similarity content, but since it is coupled to other relationships it is constantly influenced by their transformations and consequently pulled apart to some degree. In this way, some relationships become more unreconciled, which we can also describe as being more tense, due to the higher degree of dissimilarity, while others get more reconciled. The tenser relationships have a higher potential to influence the reconciled relationships, and this potential is what we regard as the energy of the physical system.

In order for this mechanism to work, there is one more property that the representational relationships must possess—they must have an individuality which remains unchanged within some limits of transformation of the representation. If that was not the case, any minute transformation would be perceived as a different physical object and there would be no persisting representational patterns which would constitute continuously existing physical objects like particles or aggregations of particles. In that case the universe would be a mess of indiscriminate activity. The representations, however, are able to preserve their identity within some limits of transformation due to the fact that they have internal configurations. When these configurations change a little bit, they still may preserve the qualitative character of their interaction (e.g. repelling or attracting each other), which endows them with an identity. They act in the same way in similar circumstances, but the resulting action differs in the degree to which they affect each other's state. For example, when two electrons collide they always repel each other, but they may do this with different strength depending on the speed with which they move. In every case the representations constituting the particles transform, but depending on whether their internal configuration remains qualitatively the same or not, we perceive them as retaining their individuality or transforming into new particles.

A representational transformation with retained internal configurations is what we regard as interacting particles with different energies, while changed internal configurations is what we regard as creation or annihilation of particles. So, energy turns out to be an abstract idea which we use to describe the potential of one representational configuration to act on another one. We regard it as an intrinsic property of one particle, which is equivalent conceptually to the one side of a representational relationship, but this is a physical impossibility! A representation always has two sides, and the illusion of thinking of one side as fixed and the other one as variable is a convenient idealization which allows physicists to develop theories and a conceptual framework of physical entities with fixed properties, e.g., 'the mass of the electron' or 'the strength of a charge', but in reality these are abstract idealized properties and every representational configuration has its own character in its relation with the environment. There are islands of stability in the configuration space where interactions produce qualitatively the same outcome, and this allows us to think of a number of qualitatively similar representational transformations as actions of one single entity and to measure its average properties.

Now that we established that particles and the associated property of energy ascribed to them are somewhat of an idealization in our picture of the physical reality cast in terms of representations, we can try to understand the two types of energy—kinetic and potential—in terms of representations, and how energy relates to other physical properties like mass and time.

Kinetic energy, as we stated, is related to the movement of a particle. A moving particle has higher kinetic energy than a particle at rest, and the faster the movement, the higher its kinetic energy becomes. In terms of properties of the representations, this means that the more one representation is transforming, the higher is its ability to influence the states of other representations that are coupled with it. To see why this must be the case consider what will happen if the rate of the transformation of the representation slows down. We saw that faster transformation is brought about by a higher degree of dissimilarity inside the representation, i.e., when it is less reconciled. Then the different elements making up the representation are in more disagreement and they make bigger jumps from one state to another in order to get to a reconciled state, which is the natural tendency in any representation. The bigger jumps constitute bigger movements in space, and also bigger influences on other representational states, which is equivalent to a particle with higher energy. If they manage to achieve a more reconciled state, then the discrepancies among the elements will be less and the transformation will be less active. The slower rate of transformation means also slower movement in state space.

The other consequence of the higher degree of dissimilarity is that the likelihood of interaction between more dissimilar representations is lower, and thus higher kinetic energy, or faster movement in space, translates into lower probability of interaction with bulk matter and therefore higher penetrating power of particles with higher energy (e.g., X-rays). It also means shorter wavelength, since the wavelength is the mathematical description of the extent of the particle in space, and a particle with lower likelihood for interaction would appear smaller due to the likelihood dropping down faster in its periphery. Given that we detect and measure properties of a particle through its interactions with other particles, the lower likelihood of interaction effectively shrinks the region of state space where it interacts readily with other particles, and this leads us to perceive it as smaller in size.

So, faster movement in space requires a less reconciled internal state, which in turn represents bigger internal shuffling of the elements of the representation and in effect bigger, though less likely, impact on other representational states. There is one more thing to consider in this picture, however. Since all relations (i.e., interactions) between particles are representational in nature, if one representation becomes more similar to another one, and therefore more reconciled, then at the same time it would become less similar with respect to many other representations (basically, all those moving in the direction of space opposite to the direction of its movement). With some of those representations the original one will be in an interactional relation, so it will tend to affect their states to a higher degree. Thus, we see that when one interactional relation becomes more reconciled, other ones must get less reconciled. The total degree of reconciliation of representations remains the same, at least at the level of physical systems much smaller than the whole universe (we will discuss the totality of all representational relations making up the universe in Chapter 3). This finding corresponds to the familiar physical law of conservation of energy, and in our picture of physical reality it is due to the fact that all interactions are representational in nature, therefore no representational elements are created or disappear.

Regarding potential energy, it is easy to see in our picture of physical reality in terms of representations why states that do not entail movement in space have the potential ability to influence other states, which we collectively refer to as potential energy. According to our earlier discussion, representations that do not change their position in state space are necessarily self-representations. This is the only way a representation can remain in a stable, unchanging state. Each self-representation has its corresponding internal configuration of the constituent elements of the representational state. As we saw, each such configuration corresponds to one type of particle, and any transformation of the configuration to a new one constitutes a particle transformation from one type to another.

As long as the internal configuration remains unchanged during the constant process of interaction of the global representation of a particle with its environment, there is no change in the particle's mass and therefore in its potential energy locked up in its mass. However, if the self-representation snaps to a new configuration (which we described as a manifestation of the weak force), then the coupling strength of the self-representation will change in the general case, but so will change the coupling of the self-representation with the other representational elements of the particle and its environment. So, any transformation of a particle from one type to another will be accompanied with a change in the mass of the particle as well as a change in the coupling strength of the particle with other particles, i.e., the perceived energy of the particle. According to the law of conservation of energy (or representations, as we stated it), nothing is lost, and some of the representational elements from the self-representation (corresponding to its mass in physical terms) are transformed into representational elements of the particle with its environment (corresponding to energy of the particle in physical terms). This is known to physicists as the equivalence of mass and energy, expressed by the famous equation $E = mc^2$. In terms of representations, it is simply the 'locking' or 'unlocking' of part of the representation in the self-representational core that gives the identity and the mass of a particle.

Within the framework of interacting representations that we developed until now mass and energy are naturally related and can transform into each other, as it is the case in traditional physics theory, but in physics energy is also related to the notion of time through the Noether's theorem. The theorem states that any invariance (or, in other words, symmetry) of a physical parameter gives rise to a conservation law regarding another, related physical parameter. For example, the invariance of the physical events in time (i.e., it does not matter when a physical process occurs, it always has the same outcome given the same initial conditions) entails a conservation law with respect to the total energy of the system, i.e., no energy is created or disappears. Other examples are the pairing of translational invariance in space (i.e., it does not matter where the physical process occurs) with conservation of linear momentum, and the pairing of rotational invariance in space with angular momentum.

So, is there a natural interpretation of the relation between energy and time established by Noether's theorem in terms of our picture of interacting representations? In fact, we already answered this question a few paragraphs earlier, when we concluded that the total degree of reconciliation of representations remains the same for a closed system. To state it again in more detail, when one representational element gets more similar to another, at the same time it becomes less similar to other elements with which it shared a greater similarity before. So, any transformation of a representation has both a gain component in terms of similarity to other representations and an exactly opposite loss component, so the net gain or loss is zero—no representations are created or destroyed.

But why is it precisely time invariance that produces this conservation law? Earlier we stated also that given the same circumstances, the representations always behave in exactly the same way, i.e., their behavior is completely deterministic. That is precisely the same as time invariance—if a system behaves always in the same way, then it does not matter when an event occurs, the outcome will be the same in all cases. So, time invariance is the same requirement as deterministic behavior, and deterministic behavior requires that the same differences between representations have the same effect on their behavior. In a closed system there will be no net gain or loss of differences, just redistribution of differences among representations, which means that the potential for action resulting from these differences, which we call the total energy of the system, is a conserved quantity.

There is one more way that energy and time are related in the traditional framework of physics—through the uncertainty principle regarding those two physical parameters. In order to understand the uncertainty relation we need to have some understanding of the physical concept of time, and this is what will be discussed in the next section, while the uncertainty principle will be discussed later on in connection with the notion of entanglement.

What is Time

We already talked about how the concept of energy is based on the idea of differences between representations which exert 'pressure' on them to transform. The vague notion of 'pressure' is what gives rise to energy, or the potential for action, but the very idea of transformations of representations due to differences between them is even more fundamental and contains in itself the fundamental concept of time which cannot be reduced to any other concept. Thus, in our ontology of transforming representations there are two most fundamental concepts—that of a representation, which is the basic 'substance' that exists, and that of a transformation, which is a process, not an entity, and thus contains in itself the notion of time. This concept of time, however, is a primitive, singular construct without features, which simply denotes the fact that different dissimilarities change representations in different ways. The possibility of different changes means that when some of them are complete, other may still be under way, so the transformation processes have different durations. In this way differences translate into processes with different durations and time as a physical parameter of the collective behavior of a system of representations. The exact rules governing the transformation processes and their durations give character to the concept of time and endow it with properties, but in our conceptual system there is always an irreducible element that denotes flow, or duration, or transformation, which is temporal in nature.

So, the primitive, fundamental concept of time is irreducible to other concepts and needs to be taken as a postulate without an explanation of why it exists, while the properties of the flow of time in a physical system can be derived from other properties of the system. One example is the relation between time and space, which we touched upon already in the section on physical and state space. We stated already that space is an epiphenomenon due to the regularities in the transformations of interacting representations and that it has some properties of state space—the further apart two objects are, the larger is the difference in their internal states, which is a property of state space. On the other hand, we just stipulated that time differentials are based on internal state differences—the larger is the dissimilarity between two states, the longer it takes for them to interact and to change to a new pair of states. In this way, both space intervals and time intervals depend on the degree of dissimilarity between states, making space and time related concepts. This is the case also in the theory of relativity and physicists talk actually about spacetime as a singular concept.

The flow of time in a physical system is also not a fixed property, but it can change depending on the physical processes in the system. We discussed already the phenomenon of time dilation in physical systems moving at speeds close to the speed of light relative to other physical systems in a stationary frame of reference. This phenomenon is well established in the theory of relativity and has also been verified experimentally. We explained it in terms of additional transformations in the state of moving particles which add up to the transformational process in a stationary reference frame. In light of our postulate that the flow of time depends on the degree of dissimilarity between particle states, we can add that the transformation of the representational state due to the movement of a system of particles relative to another system appears slower because the representational relations among the particles in the moving system appear to have an additional transformational component from the point of view of the stationary system, which is due to their relative movement. This additional component makes the dissimilarities between the states appear larger and subsequently it takes longer for the transformational processes to take place. This phenomenon is symmetrical—the transformational processes in the stationary system also appear to go more slowly from the point of view of the moving system.

This observation leads to the so called 'paradox of the twins'—what happens if one physical system, say, a space ship carrying one of two twins, travels with a speed close to the speed of light and returns to the original point of departure after some time, where the other twin is waiting? According to the symmetry principle of time dilation, the twin in each of the reference frames should see the other one age more slowly, which is a logical paradox! The paradox, however, is resolved by the fact that the two inertial frames are not equal—one of them experiences acceleration and deceleration forces, while the other one doesn't. In terms of the representational states of the particles in each reference frame, those in the moving (accelerating and decelerating) reference frame acquire an additional transformational component with respect to the particles in the stationary reference frame which persists throughout the whole journey. This is what causes the delay of the physical processes in the moving reference frame and effectively the time dilation there. On the other hand, the interactions between the particles (including those making up the body of the twin) in the stationary reference frame do not acquire an additional transformational component and the speed of the physical processes (including the aging of the body) remains the same. So, the situation is not entirely symmetrical, although if the twin in the space ship could observe what is going on in the stationary reference frame, he or she would see the physical processes time dilated on the journey outwards, but they would appear sped up at a higher rate on the inward journey, resulting in a net time compression from the point of view of the moving space ship. As a result, the twin in the stationary reference frame would age more than the twin in the space ship and will be older when they meet again.

It should be noted that it is possible to change the flow of time only in one way—to slow it down through faster movement, but it is not possible to speed it up relative to a stationary reference frame. Also, it is not possible to change the direction of time, i.e., the order of the physical events and their causality. This is a logical impossibility, and everything in the universe, as we can see, has a very logical and consistent behavior.

Also, a minor point is that in a closed system only a relatively small part of the system can be propelled to relativistic speeds while most of the particles in the system have to remain in the stationary reference frame. This is so because acceleration requires expenditure of energy and any closed physical system has limited resources of energy, most of which are used for maintaining the usual functioning of the system, such as maintaining life on earth, and only a fraction of the energy budget can be used for propelling a physical object, like the space ship, to high velocities without compromising the normal functionality of the system. We are capable nowadays of propelling single particles in particle accelerators to speeds very close to the speed limit of light, but doing this feat with any macroscopic object is beyond our capabilities.

With the preceding discussion of space, mass, forces, energy and time we have significantly improved our picture of the processes at work in the subatomic world of particles and now we can examine in more detail the process of transformation from one coordinated particle state to another. As we will see in the next section, this peculiarity in the interaction between particles produces the granularity of their properties and the properties of the interaction processes and is dubbed by physicists as 'quantum', giving the name of the whole branch of physics dealing with the subatomic world. It was one of the most surprising discoveries in modern physics which overturned the old view of the world and brought about a paradigm shift in the entire body of physical knowledge.

Quantization of Properties

In the preceding discussion of forces and energy we stated that the representations, due to their very nature, strive to get to a more reconciled state, which in effect produces something like a 'tension' in their relationships, giving rise to forces and energy. The less reconciled representations exert more 'tension', or influence, on the more reconciled ones. We stated also that the representations can differ in their internal structure, and the different possible configurations of the internal structure appear to us as different particles. Taken together, these two facts yield the picture of the representational elements of a particle going through phases of relatively high degree of reconciliation among them punctuated by sharp transitions to a new configuration with a relatively high degree of reconciliation. These are the ubiquitous 'quantum jumps' that govern the behavior of particles according to traditional physics.

The quantum jumps involve sharp transitions in the values of a number of physical parameters. These could be position in space, energy and momentum, the time of interaction with another particle, the spin of a particle measured at a given orientation in space, etc. This list involves transitions that retain the core properties of a particle, meaning that the particle retains its identity. It is possible, however, that a particle switches to a new internal configuration, which is also in essence a quantum jump, but in this case we talk about creation and annihilation of particles according to certain laws of conservation of physical properties. When thinking in terms of representations these two cases are not very different, but in the classical ontology of physics where the identity of the particles is at the most fundamental level of understanding of physical reality, creation and annihilation of particles is regarded as a different phenomenon than the quantum jumps in the properties of continuously existing particles. More advanced theories, like quantum field theory and string theory, do have a more fundamental level where creation and annihilation are regarded as reconfigurations of some more fundamental entity (the field or the strings), which is akin to our picture of transforming representational configurations.

The remarkable property of the behavior of particles to transition sharply from one stable state to another is responsible for many physical phenomena on the mesoscopic and macroscopic scale and defines the properties of various substances and material bodies, leading to the phenomenology of the macroscopic world as we know it. It is truly fundamental to the existence of material bodies and life in its present form.

One of the first problems resolved by the assumption that matter comes in small units, or 'quanta', was the photoelectric effect[2]. There, the sharp transition from emission of electrons to no emission when the energy of the light beam responsible for ejecting the electrons from a metal surface was lowered suggested that the light came in packets which had enough energy to kick an electron off the surface only above a certain

---

[2] Einstein was the first to propose a quantum explanation of the photoelectric effect in a paper in 1905, for which he got a Nobel prize about two decades later.

energy threshold. Later on, the jumps from one state to another and the associated inability of the particle to exist in intermediate states led to the practical exploitation of many other phenomena, like electrical current gates in semiconductors, which are the basis for the operations of any computer, the scanning tunneling microscope, where electrons perform jumps across an energy barrier whose properties depend on the positions and type of particles on a surface, allowing us to make an image of the surface at a single-atom resolution, etc.

More fundamentally, though, the quantum jumps are an expression of the propensity of particles to have stable states characterized by an energy minimum and to leave that state only when they receive a 'kick' from outside. This characteristic of their behavior has a natural explanation within the framework of our picture of interacting representations. We said that the particles strive to achieve a reconciled state with each other. If they happen to be in it (which is almost never the case in reality), they have no reason to abandon it unless they are not compelled to do so by influences from the environment. The state of maximum reconciliation is also a state with minimum energy, since, as we posited, states of less reconciliation exert more pressure on states with higher reconciliation. A state of complete reconciliation would mean that the representations are completely mutually reflective and they would transform in synchrony perpetually without any interference from or to the environment. This would be a state of perpetual constant motion of the particles without any changes in speed or direction. Such a state may seem very exotic, but it actually exists as a physical phenomenon and it is called superconductivity. In this state electrons move unimpeded in certain materials cooled to a very low temperature below certain threshold, and once the current flow is set in motion it can persist practically indefinitely.

In most circumstances, however, the particles are constantly influenced by other particles from their environment and they change their states or motion patterns in abrupt jumps. Let us focus on the detailed proceedings of a single jump. The jump is a transition form one reconciled state to another with an intermediate phase of unreconciled representational relationship. This is true both for the representational complex making up one particle and for the representational relation between two particles. They both can undergo a transition from one reconciled state to another. In the brief period of time when the representations are unreconciled we can say that the representational relationship actually does not exist, i.e., the representation itself disappears and a new one is created immediately after that. Since we imagine representations as information-bearing structures, we can think of the creation of the new representation as an act of 'learning' in some sense, i.e., acquiring new information, and the dissolution of the old representation as an act of 'forgetting' or 'unlearning' of the old information. This is useful because it is analogous to the real processes of learning and forgetting related to the conscious mental state which will be discussed in the next chapter.

So, to recapitulate, the acts of 'learning' and 'forgetting' occur because representations constantly transform and strive to achieve a state of higher reconciliation. The transformation of the representations is a fundamental idea which needs to be taken for granted and cannot be reduced to more fundamental postulates. This is essentially the same idea of different dissimilarities producing different effects which we discussed in the previous section on the nature of time. It is impossible for the human mind to imagine objects without space and actions without time, so some primitive concepts of time and space need to be assumed as axioms in our picture of physical reality.

Thus, the quantum jumps are a consequence of the basic capabilities of representations to transform (which is the temporal side of the concept of representation) and to have structure due to the different degrees of reconciliation between the elements of a representation (which is the spatial side of the concept of representation). The jump within a single representation making up a particle can be imagined as a reconfiguration of its internal structure. In other words, the representational elements making up the representation rearrange and change their shapes to some degree in a coordinated way. Change in one part requires changes in the other parts because a representational relation has two sides by its nature and if one of them changes then the other must change also. In a network of tightly coupled representational relations a change in one place leads to a change in the whole system.

This is true for relations between two representational systems (i.e., particles) as well—if one side changes, the other one must change correspondingly. Thus, if one representational system becomes more reconciled due to a reconfigurational jump in its internal structure, another one from those that are in representational relations with it must become less reconciled. This is so because in the previous state the existence of the representational relationship between the two systems (particles) meant that the element of the relationship on each side was equally reconciled with the rest of its corresponding system, so if one element becomes more reconciled with the rest of its system in the new configuration, then the corresponding element on the other side, due to the representational coupling, also would become more reconciled with the foreign system, which in effect means being less reconciled with its native system. Since it is part of its native system and not the foreign one, the higher degree of potential reconciliation with the foreign system does not have any physical significance, while the lower degree of reconciliation with its native system does have an effect on it and exerts pressure on the other elements to transform.

This is basically the mechanism of exchange of particles of force which we discussed earlier. We stated that what looks like a particle-carrier of force leaving one material particle and getting absorbed later on by another material particle can be thought of as a coordinated change in the states of the two material particles, rendering the particle-carrier of force a status of an epiphenomenon. This is the case with the photon which mediates the electromagnetic force, the gluons which mediate the strong force and the W and Z bosons which mediate the weak force. They all represent coordinated quantum jumps in the internal states of mass-possessing particles with a time delay between the jumps equal to the distance in space between the interacting particles divided by the speed of light. This time asynchrony is due to the fact that one particle becomes more reconciled and thus in effect it looses some tension in its internal structure (i.e., energy), while the other particle in the interactional relationship becomes less reconciled and thus gains tension and potential for action (energy as well). The higher degree of irreconciliation, however, means that the representational structure needs more time to settle in a reconciled state (which is of higher tension), and this produces the time asynchrony between the realization of the reconciled states of the two particles. This looks like a particle of force leaving the first material particle and arriving at the second material particle after a time delay precisely equal to the distance between them divided by the speed of light. The fact that the particles of force are moving always with the speed of light stems from the fact that space is an epiphenomenon as well, which we can think of as state space, i.e., distance in space is proportional to the degree of difference in the internal states of the particles. Thus, the time asynchrony between the reconciliation events of the two particles is precisely proportional to the distance in space between them, since it also depends on the degree of difference of the internal states of the particles.

There is one interesting corollary related to the behavior of particles due to the fact that they change their states through quantum jumps from one reconciled state to another. It follows that we can observe the particle only in its reconciled state but not in the unreconciled transitional period because we can judge the properties of the particle only based on its effects on other particles, but the effects on the other particles are manifest only when the transition to the new reconciled state is complete. That is the reason why all particles of the same kind (e.g., electrons, protons, etc.) look exactly the same and their core properties like charge, mass (at a given kinetic energy level), spin, etc. can be measured to a high degree of accuracy. Given an internal configuration of the representation, a particle behaves in exactly the same way in the same circumstances, which endows it with well-defined properties. Some of these properties are invariant under certain transformations of the internal state and therefore the measured value of the property remains constant within that set of transformations. For example, the charge of the electron is always—1 and its spin is always ½.

The concept of measurement and the role of the observer have the status of a deep-seated controversy in quantum physics and are closely associated with the question of what causes quantum systems to decohere, i.e., the quantum states of the individual particles to decouple from each other and to evolve in separate histories. In terms of our picture of particles as representations we can answer this question through the observation that particles would evolve in a shared state as long as they are not subject to uncoordinated quantum jumps. In this case the evolution of their states would be coordinated and therefore their states would remain coupled, but we should note right away that this requires some quite unusual circumstances from the point of view of the typical physical conditions that we encounter in everyday life. Typically, a particle is in a representational relation with many other particles, and if its state gets coupled significantly with another one of them, it would not be able to remain for long in that state due to the transformational pressure of all other particles in its environment. This could only happen in special circumstances, like very low temperature or special arrangements of the particles, as it is the case in certain materials which we described in relation to the phenomenon of superconductivity. The prolonged shared states of two particles underlie also the concept of entanglement, which is also well established in physics and which will be discussed in a later section.

The idea of gradual decoherence of the coupling of the quantum states leading to a sharp transition to a new reconciled state has lead to the hypothesis of the existence of a phenomenon called the 'quantum Zeno effect'. The question is: what would happen if we subject a quantum system to a rapid sequence of measurements? Each measurement is a process which puts the system in a well-defined state (a reconciled state). This could happen by letting the system (e.g., an electron) interact with high-energy particles (other electrons) in a particular prepared state, which in effect would constitute a measurement of its position in space (we can infer the position from the pattern of scattering of the electrons). In this way, each measurement would in effect 'reset' the quantum decoherence process and make it start from a highly reconciled state again and again. If we do the measurements rapidly enough, the system (the electron) will not have time to decohere sufficiently in order to undergo a quantum jump to a new state and its old state would in effect be 'frozen' in time. Every measurement will produce the same results. This is the quantum Zeno effect and it has been indeed observed experimentally.

In conclusion, we can say that the quantum jumps are a prominent characteristic of the behavior of matter which manifests itself everywhere and all the time. It is pertinent to all particles of matter, since all of them have quantum states, and it is even analogous to the behavior of the conscious mental state[3], which can be regarded as a macroscopic quantum-like aggregate state of multiple individual particles, as we will see in the next chapter. That is why it is important to get familiar with the quantum properties of the particles' states at the microscopic level.

Another very important idea in modern physics is closely related to the quantum properties of the particles' states discussed so far. This is the

---

[3]   The prominent Danish physicist and one of the founders of quantum mechanics Niels Bohr also noticed this analogy and wrote: "…the apparent contrast between the continuous outward flow of associative thinking and the preservation of the unity of personality exhibits a suggestive analogy with the relation between the wave description of the motions of material particles, governed by the superposition principle, and their indestructible individuality. The unavoidable influence on atomic phenomena caused by observing them here corresponds to the well-known change of the tinge of the psychological experiences which accompanies any direction of the attention to one of their various elements." (p. 99-100 of Niels Bohr, *Atomic Theory and the Description of Nature: Four Essays with an Introductory Survey*, Cambridge University Press, Reissue edition – 16 Jun 2011)

idea that the state of a particle at a given location in space can be regarded as the sum over the state transformations along all possible trajectories this particle can follow in order to get to this location from an earlier one[4]. This view on particle motion is very helpful in understanding the double-slit experiment discussed in the beginning of the chapter and it has its counterpart in terms of our picture of particles as representations. We will see how it can be explained in terms of representations in the next section.

Sums over Histories and Virtual Particles

The sum over histories approach offers an alternative explanation of the wave-like interference phenomena associated with light and particles of matter, like the ones we saw in the double-slit experiment. Instead of thinking of light as made up of waves, it proposes that light can be made up of point-like particles, but they would need to travel simultaneously along all possible trajectories from the point of origin of the light to the point of its absorption. Since light has a phase that depends on the distance from its source, the different lengths of the possible trajectories cause all those particles to arrive at the destination points with different phases, thus cancelling out partially and creating an interference pattern. This explanation turns out to be mathematically equivalent to thinking of light as waves that propagate through empty space.

This result struck the physics community because it appeared to resolve to some extent the entrenched controversy regarding the nature of light—is it particles or waves. It turned out that light can be regarded only as particles but at the cost of allowing them to travel along all possible routes at arbitrary speeds, even in excess of the speed of light. Such a proposition seems again a bit difficult to accept, so can we find a corresponding interpretation of this picture in our conceptual framework of interacting representations?

The key to understanding the sum over histories picture in terms of representations is to think of space as state space. Then, a transition from one point to another along different routes would correspond to a transformation from one state to another through different intermediate

---

[4] This idea was originally proposed by Richard Feynman when he was still a graduate student and received high acclaim in the physics community.

stages. It means that in the process of switching from one internal configuration to another inside the particle, the different elements of the representation have individual contributions to the transformation, and the contributions can be stronger or weaker, creating different trajectories in state space. Each individual particle at a given moment in its existence can take any trajectory, i.e., can transform in any of the possible ways, but the different trajectories have different likelihoods, so the average case is the trajectory along the shortest path, which is what we typically consider as the trajectory of the particle (a straight line in empty space).

The idea of contributions from the elements of a representation is analogous to another picture—that of contributions of different concepts to the evolution of the mental state—which we will discuss in detail later on, but it is useful to invoke it at this stage in order to provide some intuition on the analogy between particle states and mental states. This analogy allows us to bridge the two conceptual systems—the one describing matter (physical particles) and the one describing the mind (mental states), and to apply concepts native to one of them to the other one. So, instead of picturing the particle travelling along possible trajectories simultaneously, we can think of the particle's state being influenced by many other potential states of its internal configuration, which do not get fully actualized but only partially, in proportion of their likelihood, and its internal configuration being in something like a superposition of these internal configurational states and transforming as a superposition of states until its final definite state. This way of thinking is closely analogous to the way we make a decision, taking into account multiple factors in an instant and arriving at a conclusion that is the compromise of all of them.

The sum over histories approach can be applied not only to light, i.e., exchange of a photon between two particles interacting through the electromagnetic force, but it can be generalized to any particle interaction, including those that involve particle transformations (creation and annihilation). In this general form it is expressed visually by physicists through the Feynman diagrams. They depict the individual creation and annihilation events of particles as vertices of a graph. Each possible event is assigned a likelihood according to the mathematical calculations, and the sum of the likelihoods of all possible events (which are an infinite number!) is equal to 1. Thus, each possible event has a contribution to the typical, or average, case, and the properties of the average case can be measured experimentally to a high degree of accuracy using a system of a large number of particles and applying the measurement procedure multiple times. In this way physicists have been able to verify the sum over histories idea as a valid description of the actual properties of particles.

The possible events are infinite in number because each event can be repeated again and again in one interaction process. For example, two electrons can exchange a photon and recoil off each other, or the photon they exchange can briefly turn into an electron-positron pair which annihilates and turns back into a photon that gets absorbed by the original electron, or this brief creation and annihilation process can repeat two times in the same interaction of two electrons, or indeed any number of times but with a sharply diminishing probability. Also, the electron from the electron-positron pair can briefly emit a photon which can get absorbed by the positron or, even less likely, can engender a briefly living electron-positron pair, and so on to infinity. The number of combinations involving only electrons, positrons and photons is infinite, but there are also many other possibilities involving creation and annihilation of other types of particles.

All these briefly existing particles are called 'virtual particles' because they cannot be observed experimentally, but their existence can be inferred from the properties of the interaction processes between 'real' particles in particle collision events at high speeds. The 'virtual particles' can exist only for extremely brief periods of time, not enough to go more than the span of a nucleus at the speed of light, but there is plenty of them near any 'real' particle, so they are imagined by physicists as something like a 'cloud' surrounding the particle and changing the properties of its interaction with other particles.

In our picture of representations we can think of this 'cloud' as all the potential configurations that the internal representational structure of the particle can take. Since they are not actualized, they are only 'virtual', but nevertheless they determine the actual transformation of the particle's state which is a combination of all those possibilities. Thus, a virtual particle and its trajectory translates to a transformational trajectory in state space, which in turn depends on minute differences in the elements making up the representational structure arising from some degree of irreconciliation among them. Those differences, however, cannot grow as big as to split from the particle and form a separate representation, which corresponds to the limited scope of their existence in terms of space and time.

There is one more difference between the traditional way of picturing particles by physicists and our approach to thinking of particles as representations. Elementary particles are thought of in the framework of modern physical theory as point-like objects, and the cloud of virtual particles around them is what endows them with extension and shape. In the framework of particles as representations, however, a point-like object cannot have any internal structure and therefore would not be able to transform and to act. To see why this is the case we need to remind ourselves about our conclusions regarding the nature of space. We saw that it is an epiphenomenon which we need to think of as state space, i.e., every point in space representing a different state of the object occupying it. Now, if the object is truly point-like, it would have a precisely defined state which would have no internal reason for changing, i.e., it would be 'frozen' in time and not transforming. In order to be able to transform, an object would need to have some internal structure, but this entails some differences in the elements making up the internal structure and therefore different locations in state space. So, a particle with the ability to transform will necessarily have some extension in space!

When a particle interacts with another particle, it always acts as a unit and it retains its identity during the interaction process. This is the reason why we can use the simplification of thinking of it as a point-like object in physical theories, since conceptually we do not care (and maybe do not know) what are the individual contributions of each element of the representational structure; we only care about the final outcome of the interaction process. Since the representations act the same way in similar circumstances, we can ignore the internal details and think of the particle as a unit endowed with properties, disregarding its internal structure. In this way we in effect disregard also its extension in space, replacing it with forces that act at a distance, but this is only an idealization. If we want to be faithfully reflecting reality, we need to think of particles as possessing extension in space if we accept the fact that space is an epiphenomenon.

The point-like nature of the elementary particles like the electron is inferred by physicists from the observation of a diffraction pattern when a crystal made up of atoms of the same kind is bombarded with electrons. From this pattern it is possible to reconstruct back the shape of the 'cloud' (or the orbitals in technical language) where the electrons that are bound to the atomic nuclei reside. These orbitals have a specific shape predicted by theoretical calculations and the electrons are indeed found to move inside such three-dimensional regions on average. Since the deflection of any individual electron can be measured at an arbitrary distance from the atom, the location of the collision can be extrapolated back with an arbitrary precision, which makes it appear as if the collision happened indeed at a very well defined point in space. This provides the intuition of the physicists' picture of the electron as a point-like particle, whose true point-like nature is obscured by the veil of virtual particles surrounding it. However, in terms of representations in state space it is better to think of the particles as objects with internal structure and therefore some extension in state space, although we still need to imagine them as 'cloudy' objects due to the uncertainty in our knowledge of their internal structure.

The shapes of the orbitals, which we picture as clouds around the nucleus, represent the probability of finding a particle at a certain location in space (in terms of the measurement procedure we just discussed). So, the clouds are actually probability distributions in space for finding a particle, or in other words for the frequency with which it would interact with another particle, since this is the only way we can 'find' that a particle is there. Those probability distributions obey the so-called Schrödinger equation, which is a rule that defines mathematically a curve with a certain property of its curvature, thus setting it apart from any other curves that do not conform to this rule. This rule has been widely used in calculating different properties of the particles and their quantum mechanical behavior, so it is considered the cornerstone of quantum mechanics. It is less clear, however, as to why this is the rule that particles obey and how it can be derived from more fundamental principles. It happens to have a natural explanation in terms of representations in state space, and this will be the subject of the next section.

Wavefunctions and the Schrödinger Equation

The wavefunction is a mathematical device for calculating different properties of a particle or a system of particles, like its energy or probability of occupying a certain position in space. The probability distribution is given by the modulus square of the wavefunction values, so the physicists consider the values of the wavefunction itself to have no physical meaning on their own—only the modulus squares of those values have a physical meaning. More so, because the wavefunction has complex values but probability distributions can have only nonnegative values.

Just to remind ourselves of the basic mathematical ideas involved in the Schrödinger equation, a mathematical function is a mapping between a set of numbers represented by the x coordinate and another set of numbers represented by the graph of the function, whose values are defined as the projections of the graph on the y coordinate. In the simplest case of a one-dimensional wavefunction, the x coordinate represents position in space, while the y coordinate represents the probability of finding the particle at the corresponding x location. Since the wavefunction has complex values, it is represented by two graphs—one for the real part of the number and another one for the imaginary part. The modulus square can have only real nonnegative values, so its graph is a single line. The function can have also other mathematical properties, like a first and second derivative. These are other sets of numbers associated with the x coordinate values, which represent the rate of change of the function itself or a lower-order derivative, which is also a function, at the given x value. The first derivative is the rate of change of the function itself, or in other words, how much the graph of the function curves at the given point, and the second derivative is the rate of change of the first derivative, i.e., the rate of change of the rate of change. This shows how much the curvature is changing, i.e., accelerating or decelerating, with the change of the x coordinate.

The Schrödinger equation has two versions—one which involves both time and space as parameters and one which does not involve time, but has only space as a parameter. The first one is called the time-dependent Schrödinger equation and it describes the evolution of the wavefunction with time, while the second one is called the time-independent Schrödinger equation and it describes stationary states, or standing waves, i.e., wavefunctions which are frozen in time and do not change. The second case is a physical oddity and occurs in nature only under very special conditions, but it is simpler and easier to work with, something like an idealization of the general time-dependent case. We will consider first the simpler time-independent case and then the general time-dependent one.

The special requirement that the time-independent Schrödinger equation[5] imposes on the wavefunction is that the rate of change of the rate of change of the function in space, i.e., how fast it curves, is proportional to the value of the function at that point. The simplest function with that property is the complex sine (and for that matter also the cosine) function, although there are also other functions that conform to this requirement. This is the mathematical expression of the reason why particles and light have wave-like properties, like the one we saw in the description of the double-slit experiment. The Schrödinger equation requires the probability distribution of the position of the particle in space to be wave-like, following the ups and downs of the sine wave. But why is it precisely this rule that defines the distribution of the particles in space and not some other one? Within the framework of quantum mechanics the Schrödinger equation is a postulate from which the other equations are derived, so it does not offer an answer to this question. Quantum field theory starts with different postulates—that of a field, and particles there and their wave nature are regarded as excitations of the field, but it is still not clear why they obey exactly this rule and not some other one. There is, however, a reason why particles would obey this rule if our picture of them as representations forming state space is correct.

To see how this works, we only need to observe that the constantly changing curvature of the sine function makes each segment of the graph, no matter how small or how big (within one cycle of the function), unique in its shape. In other words, no two segments of the graph are the same within one cycle of the function. But that is exactly the definition of state space—no two regions of space are the same, since they represent different states (we saw earlier that it is not necessary to consider single points, since any particle necessarily needs to have structure and therefore

---

[5] The precise formulation of the equation is:

$$E\Psi = -\frac{\hbar^2}{2m}\frac{\partial^2 \Psi}{\partial x^2} + V\Psi$$

where Ψ is the wavefunction value, E is a constant (the energy of the particle), x is the x space coordinate, ∂ and ∂^2 denote the first and second partial derivative respectively, $V(x)$ is the external potential, $\hbar$ is Planck's constant and m is the mass of the particle.

extension). So, it turns out that the sine wave is also the simplest curve that has the property of uniqueness of each segment taken from this curve!

The sine wave describes the idealized situation of a totally free particle which can move and be anywhere in space. In reality, however, particles are constrained by the presence of other particles at some distance from them, so they can move only within a bounded region of space. In this case the *V(x)* potential is larger than zero for some regions of space and forms a potential energy barrier which the particle cannot overcome. The graph of the particle's wavefunction would still be a wavy, sine-like curve, but its undulations would be gradually diminishing the closer they get to the potential barrier. Still, the wavefunction retains the property of each segment being unique in shape.

The actual particles are also moving in space, meaning that their wavefunctions are changing with time, and these changes are governed by the time-dependent Schrödinger equation[6]. It differs from the stationary state formula in that the rate of change of the rate of change of the function in space is proportional not to the value of the function itself, but to the value of the rate of change of the function in time. In other words, the more the function curves in space, the more it changes in time. Again, the simplest function looks like a sine wave, but one that is moving to the side and spreading out with time.

So, why is it the case that the wavefunction needs to be modulus squared in order to have a physical meaning? Why isn't a function on its own specifying the probability of finding a particle at some location in space? The mathematical answer is that, in fact, in many calculations modelling interactions between particles the wavefunction is first multiplied by another wavefunction and then the modulus square of

---

[6] The formula of the time-dependent equation is:

$$i\hbar \frac{\partial \Psi}{\partial t} = -\frac{\hbar^2}{2m} \frac{\partial^2 \Psi}{\partial x^2} + V \Psi$$

where Ψ is the wavefunction value, i is the imaginary unit ($\sqrt{-1}$), x is the x space coordinate, ∂ and ∂^2 denote the first and second partial derivative respectively, $V(x)$ is the external potential, $\hbar$ is Planck's constant and m is the mass of the particle.

their product is taken to obtain the result of their interaction. In physical terms it means that the interaction happens at the abstract level of the wavefunctions and only after that it manifests itself as a probability distribution pattern. This peculiarity of particles' behavior is easy to understand within the framework of interacting representations. The wavefunction of a particle is like one of the sides of a representational relationship. It has no reality on its own, but exists and acts only in conjunction with the other side of the representational relationship, which may be part of the representational complex of the same particle or of another particle. We can think of each side in isolation, but when we consider the action of the representation, we always need to take both sides into account. Mathematically, this is equivalent to multiplying two functions, or a function with itself, which is exactly the modulus square operation.

As an aside, we can try to interpret the meaning of the simple mathematical operations to see why it is exactly multiplication that models interaction. The simplest mathematical operation is addition, and in physical terms it typically signifies two entities joining their actions exerted on some other entity. In terms of representations, it would mean two representations from the same complex acting together on a third representation (but not on each other). Multiplication is defined by repeating the addition operation multiple times, which could be due to the same element performing the action multiple times, or multiple elements of the same kind performing the same action once. In the second case, we can say also that each element of a set acts on each element of another set in the same way. In this way, multiplication models action by each element of one set on each element of another set, i.e., the fullest possible action of one entity on another, and that is the reason why it models interaction between wavefunctions. In the special case of multiplying a number with itself, it simulates interaction of an entity with (a copy of) itself. That is why the probability distribution of a particle is the modulus square of the wavefunction—we need to take into account the internal interactions of the representations making up the particle in order to model their behavior!

The wavefunctions of a particle have well-defined graphs for each well-defined quantum state. These graphs in fact represent an idealized model of a particle—assuming it is at rest at the exact moment of its most typical behavior. In reality the particle's state is in a constant flux, subject to perturbations from the environment, and it is virtually never in the exact state prescribed by the theoretical calculations, but if we make an average of all those fluctuations, either by measuring the properties of a large number of particles in the same state or by repeated measurement of the same particle, we will obtain the well-defined state prescribed by the theory.

We saw that the probability distribution of a particle being in some region of space is given by the modulus square of the wavefunction. This means that, mathematically, there are two possible ways of obtaining the same probability distribution—either by multiplying positive or by multiplying negative numbers. This is because both the square of a positive number and the square of the same negative number are equal to one and the same positive product. Physically, it means that there are two versions of the wavefunction which have exactly the same physical manifestation in terms of the behavior of the particle. The two versions are not perfect substitutes for each other, though, because some physical interactions are modelled by performing mathematical operations with the wavefunctions before taking the modulus square, and this produces different results representing different quantum states. However, there are cases when the two versions of the wavefunction act in the same way and this produces some strange effects in the behavior of the particles, like entanglement, which will be discussed later on.

The two versions of the wavefunction differ by the way their graphs can be divided into symmetric parts. In the case of positive-valued wavefunctions the graph is symmetric across the vertical axis and in the case of positive—and negative-valued wavefunctions the graph is antisymmetric, i.e., it is flipped both across the vertical and the horizontal axes (which is the same as rotation at 180°). It turns out that the symmetric wavefunctions describe only bosons, i.e., the particles-carriers of forces, while the antisymmetric wavefunctions describe only fermions, i.e., the particles that make up physical objects. Just to remind ourselves, one of the main distinctions between bosons and fermions is that the former do not prevent each other from occupying a single location in space, while the latter do. In terms of representations, we also found it helpful to think of fermions as independently existing representations, while we thought of the bosons as epiphenomena, i.e., differences between actual representations.

There is one more distinction between these two groups and it is related to the notion of spin: bosons have integer spins (1, 2, 3, 4, etc.), while fermions have half-integer spins (½, 1½, 2½, 3½, etc.). This is an important fact which has its interpretation within the framework of representations in state space and it will be the subject of the next section.

What is Spin

Originally, spin was conceived as rotation of an elementary particle like the electron around its axis, just like the Earth rotates around its North-South axis, but this idea was quickly abandoned since elementary particles are viewed as point-like objects in modern physics and therefore they cannot rotate. Still, the electron behaves as if it has some intrinsic angular momentum, i.e., a propensity to turn to one side or another, which becomes manifest when it moves through an inhomogeneous magnetic field. In such physical arrangements a beam of electrons sent through the magnetic field splits in two streams deflecting in opposite directions. The electrons from the two streams are labelled 'spin ½' and 'spin—½' respectively, in line with the theoretical prediction that the spin of the electron can have only two possible values: ½ and—½ (also called 'spin up' and 'spin down').

So, spin appears to have something to do with rotation in space, but in a rather strange way since it is a rotation of a point-like object which cannot be visualized as a spinning round ball. So, in order to understand what spin is, we need to clarify for ourselves the idea of rotation in space. Let us try to cast this question in our framework of representations in state space. First of all, we need to remind ourselves that we regard space as an epiphenomenon arising from the way representations interact. Therefore, representations do not actually have orientation in space, but they can interact in different ways which effectively results in different coordinated transformations that we can picture as movements in different directions in state space.

The simplest case of two different transformations would be if in one case the representations change their internal structure in a symmetric fashion, which would represent movement along a single dimension in space (i.e., movement along a line—either towards each other or away from each other), and in the other case the representations change their internal structures in a completely anti-symmetric (or cross-correlated) fashion, i.e., any rearrangement of two elements of the first representation will correspond to a totally different rearrangement of the corresponding elements in the second representation, which would effectively constitute movement in two different directions in space. These two directions need to be geometrically orthogonal to each other, since movement along one of them (which represents transformation of the internal structure of one representation) is completely distinct from movement along the other one (the transformation in the internal structure of the other representation). In this way, when one representation transforms, the other one transforms in a cross-correlated way, meaning that each of its elements performs a different transformation than the corresponding element in the first representation, and this effectively creates a second dimension in state space along which the second representation moves.

So, the ability of representations to undergo coordinated transformations creates movement along a single dimension in space, while the ability to undergo cross-correlated (or anti-coordinated) transformations creates movement along two dimensions in space. We can picture this as a movement of one particle along the x axis and simultaneous movement of another particle, which is in an interactional relation with the first one, along the y axis. Note that this picture matches precisely the familiar picture of the action of electromagnetism! When electrons move along a wire, they create a force that makes other electrons move orthogonally to the wire, i.e., in concentric circles around it. This is the principle on which electric motors are based, and it illustrates very well the intrinsic property of spin associated with electrons.

There is another way of schematically illustrating the aforementioned relations between representations, which we regard as the counterpart of the physical notion of spin—we can do this by drawing arrows as in Figure 9.

a) Spin ½ - possible representational relations at ½ turn

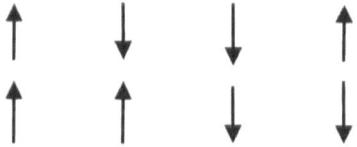

b) Spin 1 - possible representational relations at 1 turn

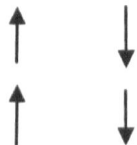

c) Spin 2 - possible representational relations at 2 turns

Figure 9: **Schematic depiction of the possible representational relations for spin-½, spin-1 and spin-2 particles.**

The spin-½ particles can be depicted as arrows pointing in orthogonal directions with respect to each other. The possible cases are shown in part a) of the figure. Physically, this case represents the elementary fermions: leptons (electrons, muons and tau plus their respective neutrinos) and quarks (making up the protons and neutrons). The spin-1 particles can be shown as arrows pointing along the same line, as shown in part b). The physical particles belonging to this case are the elementary bosons: the photon, the gluons and the W and Z bosons. The particle-carrier of the gravitational force—the still undiscovered graviton—is assigned spin 2 by the theory (part c)).

It is interesting to observe that if we define one 'turn' as a rotation at 180°, then the different configurations of the arrows are all possible ways of rotating the arrows with respect to each other a number of 'turns' corresponding to the value of the spin—½, 1 or 2. In physics, the value of the spin derives from the solutions to a mathematical equation describing the spin states, but these numbers turn out to correspond well to our picture of the spin of representations in terms of 'turns' of the arrows that symbolize representational relations. So, if the two arrows (or representations) are orthogonal to each other, meaning they have a spin-½ relation, then there are 8 possible ways in which they can arranged. In all these relations one of the arrows is at a half a turn with respect to the other arrow. The possibilities of two arrows differing by 1 turn from each other are 4, and the possibilities of two arrows differing by two turns (either two turns of the same arrow or one turn of each arrow) are 2.

It turns out that this illustration in terms of arrows matches the properties of spin in one more way. In physical theory the spin of the particles has a mathematical property called phase which is also related to the mathematical notion of rotation in phase space. Thus, the phase of the spin-½ particles needs to be rotated 720° before returning to the same quantum state, the phase of the spin-1 particles—360°, and the phase of the spin-2 particles—180°. If we assume that each arrow configuration represents mutually exclusive (and therefore orthogonal in the geometric sense) phase, then the number of arrow configurations in our schematic depiction corresponds exactly to the number of orthogonal rotations (at 90°) of the phase: 8x90° = 720° in the spin-½ case, 4x90° = 360° in the spin-1 case, and 2x90° = 180° in the spin-2 case. Although it is not clear if this analogy can be extended to other values of spins of composite particles (e.g., spin-3/2, which is the spin of the delta baryon, or spin-4, which is the spin of some resonance meson particles), it can be useful as a bridge between the conceptual framework of modern physics and our conceptual framework of representations in state space.

The main lesson, however, from our picture of spin as a property of the representations is that it can create additional dimensions in state space, thus effectively defining the dimensionality of the space. We know from experience that we live in a three-dimensional universe, and according to our conclusions this would be due to the capabilities of the representations to transform in cross-correlated, anti-symmetric ways. Since the fermions can do this, they effectively create two-dimensional regions of state space for each representational interaction and those tiny regions join to form three-dimensional space in large-scale interactions of multiple particles. The orthogonal nature of the transformation of the representations of spin-½ particles explains also their inability to occupy the same location in space: as one of the representations transforms in a way that will bring it closer to the other one, the other one transforms away from that state, so the two are constantly chasing each other but never catching up, which looks like the two particles orbiting around each other in space but never merging in a single particle.

Spin, as we said, is a property related to the electromagnetic interaction, but there are also two other interactions between fermions—the strong and the weak—and they also turn out to have a property like spin associated with them: isospin in the case of the strong interaction and weak isospin in the case of the weak interaction. They are somewhat different from spin and also from each other, but the common feature of all three is that they can be represented mathematically as rotations in some kind of abstract phase space and that they take on discrete values. The proton and the neutron are considered in this framework to be in fact one and the same particle with different values of their isospin. The basic idea which we can use to understand them, though, is the same: they represent the capabilities of the representations to transform in mutually independent ways when they engage in an interaction.

The spin of the electron can take on two values—either 'up' or 'down' (½ or —½). Although the electrons are fermions, it turns out that if they have different values of their spin they can occupy one and the same orbital around the nucleus of an atom, meaning that electrons with different spins are in a sense more compatible and interfere less with each other's presence at the same spatial location.

When moving through an inhomogeneous magnetic field, electrons with spin up get deflected in one direction, and those with spin down in the other direction relative to the lines of the magnetic field. This effect is the basis for the Stern-Gerlach experiment, named after the two physicists who performed it first. Segregating electrons with different spins into two beams turned out to be a very useful tool for studying quantum phenomena, and variations in the setup of the experiment were used to illustrate another enigmatic property of quantum systems—that of entanglement, which is the subject of the next section.

What is Entanglement of Particles' States

It is relatively easy to produce and direct a beam of electrons travelling through space—this is what happens inside the cathode ray tube (CRT) monitors and older tv sets. When such a beam is passed through the inhomogeneous magnetic field of a Stern-Gerlach apparatus, we obtain two beams—one going in one direction (let's say, up) made up exclusively of electrons with spin 'up', and the other one in the other direction (down) made up of electrons with spin 'down'. If we now place a second Stern-Gerlach magnet in the path of one of the beams, say the 'up' beam, we can check if the electrons behave the same way as before. Indeed, what we find is that if the magnet has the same orientation as the first one, then there is only one beam exiting the second apparatus, and it is deflected upwards. If, however, we rotate the second apparatus sideways (at 90°), then the beam is split in two halves again, which are labelled also spin up and spin down, but this time along another dimension—the x axis. Now, if we place a third apparatus along the path of the spin-up beam exiting the second apparatus and we rotate it back to the same orientation as the first apparatus (at—90°), then surprisingly we see again two beams—one with spin-up electrons and one with spin-down electrons. What this means is that the second apparatus must have 'erased' somehow the spin state along the z axis, which was exclusively spin-up at the entry of the apparatus, but is random at its exit.

This phenomenon is an example of the entanglement of the spin states along the three different spatial dimensions—defining the value of the spin along one of them effectively erases the values of the spin along the other two.

Things get even trickier if we add another twist to the experiment: instead of using an ordinary beam of electrons we can use electrons (or photons) produced in a single quantum event. In that case the electrons are emitted in exactly opposite directions, so we need to place two Stern-Gerlach magnets at the opposite sides of the emission chamber. Due to conservation laws, the emission event produces pairs of electrons with exactly anti-correlated spins, i.e., if one of them is spin-up in the z direction, the other one must be spin-down. This is actually what we find in the experiment when electrons are emitted one pair at a time. Although we cannot control which apparatus will detect the spin-up and which one the spin-down electron, they always show opposite results.

The tricky part comes when we rotate one of the magnets so that they are no longer aligned in their orientations. In this case the two electrons' spins will not be exactly anti-correlated, but one would expect that they would be correlated with a probability proportional to the angle of rotation. For example, if the second apparatus is at a 90° angle with respect to the first, they will be correlated with 50% chance and anti-correlated with 50% chance. If the apparatus is at another angle, the percentages should change accordingly, e.g., at a 120° angle the anti-correlation ought to be 1/3 of all observations and the correlation 2/3 of them. This, however, is not what is found experimentally! In fact, at a 120° angle the anti-correlation is 1/4 of the cases and the correlation is 3/4 of the cases. This is exactly the result predicted by quantum mechanics for a pair of particles in a shared state.

A somewhat more complex setup of the experiment involving random rotations of the magnets at 120° from each other yields the additional finding that the total number of anti-correlated cases matches the total number of correlated cases. A more complex line of reasoning shows that this is impossible if we assume that the particles have some intrinsic values of their spins that are set at the moment they fly apart from each other. This means that the particles acquire the values of their spins at the moment they interact with the magnets, which, however, can be at an arbitrary distance apart. This is a very puzzling finding since it appears to violate the speed of light limit with which any interaction can spread in space, and it has produced a lot of controversy in the attempts of physicists to find a satisfactory explanation for it.

The modern interpretation of these results is that particles located an arbitrary distance apart in space can have a shared quantum state which can acquire definite values for its quantum properties, like spin, momentum, position in space, etc., effectively making both particles acquire definite, correlated values of the same property at the same moment, but there is always an element of randomness in the fixing of the values, i.e., it is not possible to control which value will be obtained, thus making the communication of a signal using the entangled pair impossible. In this way, the principle of the limit of the speed of light for the propagation of any interaction process is not violated and there is no instantaneous action at a distance, just instantaneously set correlations.

This unusual property of the quantum systems actually has a natural explanation in terms of our framework of representations creating state space. To see how it works, we need to focus again on the very basics of the concept of a representation and how representations create state space.

As we stated earlier, the concept of representation is based on the concepts of similarity and difference—any representational relationship has both some degree of similarity and some degree of difference. Now, since space is a state space, the similarity component would entail the same location in space, while the difference component would entail a separation in space, or some extension of the representation. When the representation is unified and survives an interaction with another representation, we think of it is a unit which persists in time, i.e., a particle. Thus, a particle can be pictured as a tangled knot of multiple representational relationships, some of which are directed to each other and fall completely within the particle (the self-representational core), and others which are directed to other representational complexes belonging to other particles in the environment.

We also stated that particles interact by switching from one well-defined state to another in coordinated quantum jumps. In this way the representational relationships among them are conserved, and this allows also for representations to disengage from one particle and to engage with another particle. In a way, each such switch, or quantum jump, represents a move from a tendency to coordinate the evolution of the quantum states on both sides to a disruption in the coordinated evolution, i.e., from a drive to increase the similarity component (which, if extrapolated, would merge the two particles into one) to a restoration of the normal ratio of similarity vs. difference through the disruption of the growth of the similarity component. We can picture the similarity component as some kind of 'glue' or 'thread' that connects the different representations, while the difference component creates separation, or empty space, between them.

Now, let us consider what happens in the case when a unified quantum system splits in two parts, as in the process used in the Stern-Gerlach experiment described above. Being unified means that it has connecting 'glue' between all of its components. If it splits in two, then at least one of its self-representations needs to disengage, leaving two unconnected parts. This means that the similarity component of this representation must move elsewhere, to other representational components within the particle system. So, the result will be two separate representational components, but with some degree of mutual representation, since the total degree of similarity and difference must be conserved.

From the point of view of the environment nothing would change until an interaction process is complete resulting in a quantum jump. Each of the new representational complexes would behave as before even after the split, since the representations connected with the environment would belong either to one of the parts or the other (they don't split in halves). So, a difference in the behavior of the new particles will become evident only after a quantum jump in one of the new systems is completed, due to pressures from the environment. Before that occurs, the two particles would behave as if they are still one system, due to the fact that all external representational relations are preserved and the representational change is due only to a disengagement of an internal representational component.

In terms of the movements of the particles in state space, the two new particles would transform in a strictly symmetric way (since they are still exclusively in representational relations with each other), which means they need to move in exactly opposite directions in state space and with exactly the same speeds. This is just what is observed in the quantum events of splitting, like nuclear fission or two-photon emission. In fact, we should not think of the particles as splitting off and flying apart from a single location in space, what would be the classical depiction of the event, but as a system of representations dividing in two and transforming in two different, symmetric ways. Until it has interacted with the environment, i.e., exchanged a photon with it, it is impossible to know if the two parts would continue their evolution independently or re-merge in a single complex, so we need to think of the particle as being in a superposition of a split and a unified state. However, due to its internal evolution, the location of the particle in state space becomes more and more different for the two parts, so in effect it is in a superposition of being a one whole at the original location and being two particles in opposite locations in space. As the internal state of the particle transforms, it becomes more and more likely that it will split, and less and less likely that it will re-merge.

The instantaneous fixing of correlated values of some quantum state is a result of the complete integration of the representational information inside the particle. Although it evolves in two distinct ways, it still remains one complex where each representational element is completely representing the rest of the elements, so any change in one of them affects all others. The moment a photon is exchanged between one of the two parts and the environment, which means that a representational change is fixed and the representational complex of the particle looses some old information about the environment and gains some new information, then the evolution of the two parts splits and continues in separate ways. The reason for this is that the photon exchange occurs entirely within one of the parts, therefore the evolution of the two parts in no longer symmetric. However, the fixing of the representational transformation that constitutes the photon exchange switches the representational state of both parts to a definite, observable value at the moment of the fixing, so the measured quantum states of the two particles are correlated.

So, why is it not possible to send a signal using this internal connection between the two parts of the evolving representation? The impossibility stems from the fact that the connection exists only as long as the two parts of the representation are completely reflecting each other and their evolution is entirely symmetric. In order to send information, we need to change the state of the transmitting medium—this is what it means to transmit information. If, however, we affect the state of only one side of the entangled pair, then the link between them will break at the very moment the state of the affected side is changed, so it would be impossible to tell what the change was by measuring the properties of the other side.

It is possible to 'nudge' one of the sides only slightly, so that the entanglement connection will not break apart, but in this case some special conditions need to hold, making it again impossible to exert a causal transformation of the state of the remote side. More specifically, in order to implement a 'nudge', the two sides of the representation must be very similar in their representational content. Since they reflect their environments, we need to arrange so that the environments are very similar too. In that case the slight perturbation of the state on one of the sides can be matched by a similar perturbation in the state of the other side. That, however, would not constitute transfer of information either, because we need to pre-arrange the state of the remote environment to match the state of the proximal environment, so we would already know the arrangement of the remote environment and its new state, which does not convey any new information to us.

Suppose, however, that we make the two environments similar enough and we change the state of the proximal side just so that it transforms enough to be measurable but not enough to break the entanglement connection, resulting in a measurable change in the state of the remote side. This would indeed constitute a transfer of information. In this case, however, the transfer still will not be faster than the speed of light for the following reason: we stated that space can be regarded as state space, meaning that the space intervals reflect also differences in the internal states of the particles. The entangled pair, as we saw, is also subject to that rule, with its difference components on both sides growing more distinct as the time/space separation between them grows. So, in order for one of the sides to exert a causal transformation of the other side, it needs to go the whole way of converting the other side's internal state to the new state. This process would take more time the more different the two sides are, which precisely matches the spatial separation multiplied by the speed of light. Again, the causal transformation which transmits information occurs over a time interval that matches the time needed for a light signal to travel the distance in space that separates the two parts of the representation, making it impossible to send a signal faster than light.

What is special, and different, in the case of using entangled particles to send information, rather than sending light signals, is that in the case of entangled particles we have correlated changes in the states of two remote entities, while in the case of a light signal we have a change in the state of one entity followed by a change in the state of another entity some distance apart. The time it takes for the causal change to propagate from the first entity to the second is the same in both cases, but the mechanisms are somewhat different, due to the different representational relations between the entities and their environments in the two cases.

Besides the use of the entanglement phenomenon as a means for sending information, it has other practical applications, the most important of which at the present moment are quantum computation and quantum cryptography. The second one is again a way of sending information, but due to the properties of entanglement it is possible to devise a scheme which makes it impossible to eavesdrop on the signal, making the communication perfectly secure. The first one—quantum computing—holds an even greater promise for practical utility by making possible to speed up some computations by orders of magnitude compared to classical computations. It exploits the ability of quantum systems to interact while being in superpositions of pure quantum states (well-defined representational states, in our terminology), which effectively corresponds to multiple classical computational steps executed simultaneously. With this approach it is theoretically possible to find information in an unordered database much faster than with classical computation and also to find the factors of a large number, which is the basis for modern-day cryptography that makes communication, and therefore commercial transactions, on the internet secure.

Quantum computation might also be hugely important as a first step towards the development of artificial consciousness, but in order to understand how this might come about we need to understand what consciousness is, and this will be a subject in our discussion later on. For the moment, we will stay within the domain of modern physical theory and we will explore another important topic in quantum mechanics, closely linked with entanglement, namely Heisenberg's uncertainty principle.

Heisenberg's Uncertainty Principle

We saw that within our framework of representations in state space the phenomenon of entanglement stems from the fact that representations have components of similarity and difference, the first of which is responsible for the connection between representations, i.e., brings two representations in contact with each other, while the second one is responsible for the spatial separation between them. In entanglement, the difference component grows, while the representations retain at least some similarity component linking their states, and this is what makes the system behave as one whole even if the representations are some distance apart.

This picture of representations made up of a similarity and a difference component has one more consequence, namely, that the measurable properties of the physical systems (particles) are not completely independent, and fixing one of them might influence the measured value of another one, just like in entanglement fixing a physical property of one system affects its counterpart in another system. The difference between the two cases is that in the former the correlations are between two properties of the same system, and we think of it as uncertainty in the measured values of the two properties, while in the latter the correlations are between the same property in two different systems, and we think of them as being entangled.

The prototypical example of two properties that are not independent, and actually the first such pair of properties that was found to obey the uncertainty relation, is the mutual dependence of the measurements of position in space and the momentum of a particle. At any given moment we can measure either the position or the momentum, but not both simultaneously. If we try to come up with some clever setup that would allow us to measure both properties at the same time, we come to a fundamental limit of how precisely we can measure (and know) both properties simultaneously—if we increase the precision in the measurement of one of them, then the precision of the other one deteriorates. This is the essence of the uncertainty principle—it stipulates a physical law in the form of a mathematical equation that sets a limit on the precision of the simultaneous measurement of a pair of properties. It works also for entangled particles—if we measure one of the properties on one of the particles, we are limited in the precision with which we can measure the other property on the other particle, which prevents us from deducing the value of the second property for the first particle.

So, why are exactly position and momentum linked with an uncertainty relation? Why is it not possible to define both of them with an arbitrary precision simultaneously? There is an explanation for this phenomenon in terms of the classical concepts of particles moving and interacting in empty space, but it arises also in our picture of representations creating state space. First, let us consider the classical explanation.

To know the position of a particle we need to make it interact with one or more other particles, whose movements we can trace, and from those movements we can conclude what was the position of the measured particle. The precision of the measurement is limited by the size of the particles that are used to make the measurement, and their size is inversely proportional to their energy, due to the fact that more energetic photons or electrons have shorter wavelength, and therefore smaller size. So, in order to measure the position with higher precision we need to make the particle interact with another particle with higher energy, and this effectively increases the uncertainty in our knowledge of the momentum of the particle.

In order to measure the momentum with a high precision, we need to make the particle interact with other particles with a well defined momentum. However, a well-defined momentum implies well-defined kinetic energy, and this in turn requires that the wavefunction of the particle is made up of a narrow range of frequencies, effectively making its wavefunction more spread out in space. Thus, the higher precision in our knowledge of the energy and the momentum of the particle leads to a more spread-out region of probability of finding the particle at some location in space, which decreases our knowledge of the position of the particle.

Now let's see how this relation between position and momentum can be explained within our picture of representations in state space. The explanation is actually rather simple. Increasing the probability for finding a representation at a certain location in state space means that the representation should be found with high probability in one and the same state (corresponding to the same location in state space) in any potential interaction with another representation. However, the other representations can have a variety of states with different levels of energy, which means that the measured representation must have more variety in its structure in order to respond in various interactions in the same way, but at the same time it must have a high degree of similarity in its structure, in order to stay in the same state. These requirements make the internal relations between the representational elements more diverse, or in other words, increase the variety in the tensions between them. This affects both the self-reflective component, which is responsible for the mass of the particle, and also the representational components related to the other particles in the environment, effectively making the potential for action on them more diverse. This is precisely what is defined as the momentum of the particle, which is the product of its mass and kinetic energy.

If we make an experimental arrangement to measure the momentum of the particle with high precision, this means that in different measurements we will need to find the particle possessing the same product of mass and kinetic energy. Again, the only way to make the measurement is to register the consequences of the particle's interaction with other particles, and this happens via exchange of (real and 'virtual') photons. So, the momentum will be more precisely defined when the exchanged photons are carrying the same energy, more or less. This means that the tensions between the components inside the representation must be more uniform and therefore the elements making up the structure of the particle must be regularly spaced out in state space. Compared to the average case of higher variability in the structural configuration and tensions between the elements, the more regular configuration would be more spread out in space, since there will be no small gaps in the spatial structure. This means also that the particle can be found in a number of different states, corresponding to a number of locations in state space.

In this way the position of the particle in space is more variable and consequently the uncertainty in our knowledge of the position is higher.

Besides the position-momentum uncertainty relationship, there are a few others, like the energy-time uncertainty relationship, two orthogonal directions of measurement of the spin of spin-½ particles, angular position and angular momentum in the case of small angular uncertainty, and some others. They all can be regarded as a consequence of the fact that the measured properties are not defined at the most fundamental level of description of the physical systems, but epiphenomena that can be reduced to more fundamental notions, like our framework of representations forming a state space, and the link (or entanglement) between them arises because they are partially one and the same thing—if we regard space as state space, then position in that space also determines to some extent the behavior of the particle, and therefore its potential for action, which is what we consider the momentum of the particle. In the same way, the energy of the particle determines to some degree how long a process of interaction of the particle with other particles will last, since both time and energy are dependent on the degrees of difference among the structural components of the representation, as we saw in the earlier sections. Spin, as the property that gives rise to the dimensionality of space and therefore the possibility of rotation and the different orientations in space, is also related for two orthogonal directions of measurement. So, the uncertainty relations also turn out to have a meaningful explanation in the framework of representations in state space.

After this lengthy discussion of the different phenomena that make up quantum reality so unusual and unfathomable for common sense, we can now return to our initial picture of the double-slit experiment and try to imagine what is going on there using our newly acquired knowledge of the workings of the quantum world in the framework of representations and state space. What seemed to be a rather simple picture of particles that behave like interfering waves turns out to have a much richer phenomenology involving many other fundamental notions like mass, forces, energy, spin, virtual particles, entanglement, etc. As we saw, all this complexity can be reduced also to the simple picture of interacting representations, and this picture may be able to provide an even better understanding of the phenomenology of the double-slit experiment than the classical picture of interfering waves, and ultimately also of any phenomenon in the microworld.

The Double-Slit Experiment—Revisited

According to the classical description of the double-slit experiment photons of light are emitted from the source and propagate as tiny waves in empty space until they reach the barrier with the two openings. At that moment the waves squeeze through the openings and get re-emitted as two separate waves on the other side. In this way the two slits effectively become two sources of waves. On their way from the slits to the screen the two waves interfere with each other and produce a striped pattern on the flat surface of the screen.

This rather simple story is able to explain one particular phenomenon—the striped pattern formed on the screen behind the two slits, but the wave nature of light is unable to explain many other phenomena in the quantum world, so it can be regarded only as an ad hoc explanation which is useful only in some circumstances but not in others. In order to obtain a universal explanation which will be valid in all circumstances and would apply to all quantum phenomena we need an ontologically different explanational framework which does not derive from everyday experience, like classical physics, but is built upon a different fundamental notion of matter, one derived from modern quantum physics. As we saw in the preceding section, the role of this fundamental notion can be played by the existing concept of representation.

In the framework of representations forming state space, instead of picturing photons propagating as tiny waves through space and interfering with each other, we need to imagine the setup of the experiment as an arrangement of representations which are related to each other, i.e., affect each others' states. For simplicity, let's imagine only a minimum number of them—one for the source, a few for the slits and a few for the screen. In effect, these are individual particles, i.e., atoms made up of protons and neutrons bound up in atomic nuclei with electrons bound around the nuclei, and each particle is in our terminology a representational structure (something like a tangled ball of representational elements).

The representational elements making up the particles are in constant flux, undergoing a constant process of transformation. Some of them are related to elements within the particle, forming a self-reflective part which acts as an anchor for the state of the particle, effectively endowing it with mass, while others are directed towards other particles in the environment. In particular, they are directed towards the few particles making up the barrier with the two slits. The arrangement of the particles in the barrier is such that some of them are close to each other, while those making up the edges of the slits are at some distance apart, corresponding to about the wavelength of the photons in classical physics terms.

In terms of the picture of representations the distances between the particles are defined as degrees of difference between their states, effectively forming a state space in which they are situated, and which plays the role of empty space in classical physics. Thus, the distance from the source to the barrier in state space is large and the distance between the particles making up the barrier is small except for the distances between the particles on the edges of the slits, which is on the order of less than one cycle of transformation of the representation. The distance from the barrier to the screen is again large, on the order of many cycles of transformation of the representation. These distances correspondingly reflect the degrees of difference in the states of the representational complexes making up the particles.

The constant transformations of the representations exert pressure on the particular configurations of representational elements in each particle, and these configurations are prone to snapping into a new configurational arrangement. This is what we described as the action of the four fundamental forces on the particles. Since all representations are related to each other, the switches to new configurations are coordinated between pairs of particles, which looks like a particle-carrier of force exchanged between them. In particular, when a coordinated change in two representational elements from two different particles, each directed at the other particle, occurs, then we consider this as an exchange of a photon, i.e., an action of the electromagnetic force.

Let us consider now what happens when a representational element from the particle that constitutes the source exchanges a photon with another particle. If the other particle belongs to the barrier, we say that the photon emitted by the source was absorbed by the barrier and this event is not of great interest to us. The more interesting case is when the photon emitted by the source is absorbed by a particle in the screen. In this case, the coordinated change occurs between the particle constituting the source and one of the particles constituting the screen, but the exchange is also influenced by the particles of the barrier which are inbetween them.

In the ontology of representations the exchange of the photon between the two particles (of the source and the screen) happens through a coordinated process of transformation of the representational complexes of the two particles. This process goes through stages that would correspond to the flight of a photon along a trajectory through space between them. Since space is a kind of state space, the trajectory of the photon traverses a line in state space, which is the graphical depiction of a process of transformation, namely, the coordinated transformation of the states of the interacting particles.

In the case when there is a barrier made up of other particles inbetween the source and the screen, the trajectory of the photon can pass only through the openings in the barrier. The consequence of this arrangement for the states of the interacting particles is that they must go through a stage of a well-defined state (at the moment the photon passes through the openings) which gradually gets less defined after that. If there is a single opening in the barrier, we can picture the representational elements of the source particle, related to elements in the screen particle, as clustering to a single state/location in space at the moment of time corresponding to the distance from the source to the opening in the barrier divided by the speed of light. When there are two slits open, we need to picture the representational elements forming two clusters, each corresponding to the state of state space marked by the position of the slit. Since the two clusters are part of the unified representational complex of the particle, they are tightly coupled and interact strongly. So, after the segregation of the representational elements in two clusters they gradually dissipate and after a while relax to the original configuration of more or less uniform distribution of states.

The segregation of the representational elements in two clusters explains why the source particle prefers to interact with some of the particles in the screen but not others. As we saw in the section on the Schrödinger equation, the changing state of the particle is connected with a changing likelihood for the particle to interact with another particle, resulting in an undulating probability distribution for finding the particle at a given location in space. Thus, since space is a kind of state space, the probability of finding a particle at some point in space oscillates from high to low with distance. This phenomenon can be visualized by the familiar pattern of concentric waves spreading from the two slits and interfering with each other in the classical explanation of the double-slit experiment. So, when the representational elements form two clusters, they all are in a well-defined state and their subsequent transformation continues in synchrony (which is gradually degraded by influences from the environment). The synchrony means that all representational elements would interact with the same likelihood with another particle at a given moment in time, which can be depicted as a concentric line of uniform probability in the space behind the slit. The gradual dissipation of the well-defined state is depicted by gradually diminishing and more spread-out waves.

Since the screen behind the barrier is flat, the distance from the slit to each subsequent particle in the screen is gradually increasing, meaning that the probability for interacting with each subsequent particle is gradually changing from high to low. This is what produces the striped pattern on the screen.

The effect of the slit is to put the representational elements of the interacting particles in the same state. If there were no slit, they can remain in a diverse range of states throughout the whole process of transformation (in the terminology of the sum over histories interpretation, they can follow many different trajectories from the source to the screen), but when there is one or more narrow openings, they are forced into a uniform state. Also, the width of the slit has to be sufficiently narrow, less than the range of variability of the states of the representational elements (which corresponds to the wavelength in classical terms), for the state to become sufficiently uniform in order to synchronize enough to produce the striped pattern.

So, how can we explain the odd finding that some locations in the screen receive photons when only one of the slits is open, but cease to receive photons when the second slit is open? As we saw earlier, this fact was very difficult to explain in terms of the classical picture of photons as tiny objects moving through empty space and it gave rise to such bizarre interpretations as parallel universes and 'unreal' particles.

In fact, this phenomenon is very easy to explain in the framework of representations and state space. The cases of one open slit and two open slits are very different because the way the representational elements congregate in clusters is different. In the case of one cluster, the representational element may be related to a particular particle in the screen, resulting in an exchange of photon with it, but in the case of two clusters this may no longer be the case since the representational elements may be related to other particles and not to the particle from the first case.

The picture of particles as representation explains also the version of the experiment where the choice of whether to open the second slit or not is made after the photon is emitted from the source, i.e., while it is on its way to the barrier. According to the classical picture it is hard to imagine how the photon could find out how many slits are open and behave accordingly, but in our framework of particles as representations the representational elements are always in touch with their environment and can react to changes in it. This can happen even during the course of the coordinated transformation of two particles. For example, if at the onset of the process the two slits are open, the transformation will evolve towards a state of two clusters of representational elements, but as one of the slits becomes closed, the particles occupying the former empty space of the opening exert pressure on the representational elements of the source particle forcing them to reconfigure to a single cluster corresponding to a single opening. This is how the photon can always be 'aware' of any changes along its trajectory and behave accordingly.

Until now we have discussed the phenomenology of the quantum behavior of single particles. Although this discussion is far from complete and there are literally scores of phenomena left untouched, it should suffice for building the basic imagery and conceptual understanding of the workings of the quantum realm and therefore inform our concept of matter. Developing the ability to imagine matter as representations forming state space, with all the rich phenomenology discussed so far, is a prerequisite for understanding human consciousness and ultimately the totality of all that exists, which we label 'the universe'.

From now on, we will discuss larger and larger structures of particles, starting with atoms and molecules and moving on to biological organisms and the brain, which is a very special object since it can sustain conscious processes. The brain will be the subject of Chapter 2, and in order to prepare for this discussion, we will try to understand also the process of evolution of biological organisms in light of our picture of matter as interacting representations.

Aggregations of Particles (How To Imagine Bulk Matter)

It is considered, according to modern cosmological theories, that nearly all of the long-lived particles, like the protons, neutrons and electrons, were created some time soon after the Big Bang and their number has not changed significantly since then. What has changed is only their distribution in space, which has gone from being more uniform to becoming less and less uniform under the action of gravity. So, most of the ordinary matter in the universe (the one we are familiar with and we can observe) exists in the form of large clusters of particles making up the stars and planets orbiting around them. Most of the matter of any importance to us, i.e., that which makes up our bodies or which can affect us in any way, is a vanishingly tiny fraction of all matter—it is almost exclusively the particles that make up the surface of the Earth. They make up the objects we interact with and on which we depend for survival and well-being.

Compared to the huge variation of conditions to which most matter in the universe is subjected—from the near-absolute zero temperatures of open space to the extremely strong gravity and magnetic fields of neutron stars and the unfathomable conditions inside black holes—the surface of the Earth is a rather peaceful place, with a narrow range of temperature variation and little cataclysmic activity involving displacement of large masses of materials. There is still quite a lot of activity going on in terms of rearrangements of particle structures, mostly due to the constant stream of photons from the Sun striking Earth's surface, and increasingly also due to the impact of life forms on Earth, culminating in modern-day human activities.

The particles making up the Earth and all other cosmic bodies in general are not just packed together in uniform arrays; rather, they aggregate in clusters with higher degree of attractive binding among the particles inside the cluster than between particles from the cluster and particles outside of it. The first level of clustering is that encompassing protons and neutrons, resulting in the formation of atomic nuclei. The protons and neutrons (collectively called 'nucleons') are bound together in a very tight space, in the range of a few femtometers, by the residual strong force (a.k.a. the nuclear force). This force is sort of a by-product of the strong force which binds three quarks to make a proton or a neutron. Normally, the quarks belonging to one nucleon interact only with the other quarks from the same nucleon, but if another nucleon happens to be very close to it, then occasionally the quarks may interact with a quark from the other nucleon. When this happens in quick succession, it produces this residual strong force which has the effect of keeping the two nucleons close together.

The nuclear force is extremely strong, much more than the electromagnetic force under normal conditions, like those on the surface of the Earth, and that is the reason why atomic nuclei are produced in large quantities only in the violent conditions of stars' interiors and powerful supernova explosions. On Earth they can be produced in particle accelerators but only in vanishingly small numbers. What this means is that the atomic composition of the Earth is pretty much fixed; only rearrangements of the existing structures of atoms are possible, but no transformations from one type of atom to another, i.e., one type of cluster of nucleons to another.

The next level of clustering is the atomic level. The positively charged protons inside the nuclei attract negatively charged electrons and bind together in atoms through the electromagnetic force. Since this force is weaker, it is possible for atoms to have a few electrons more or a few less, producing charged ions. Those ions (and neutrally charged atoms) in turn bind together in larger chunks to form molecules. Molecules are configurations of atoms which in conditions typical for the Earth's surface remain stable when they interact with other clusters of particles, endowing them with some permanence and therefore identity as independent objects. Since the forces that are holding them are even weaker, the clusters can break up and the remaining chunks can regroup in other clusters. These phenomena of transformations of molecules can be very diverse and very complex, and their study gave rise to the scientific discipline of chemistry.

Molecules aggregate in larger clusters, making up the macroscopic objects around us. There are two basic categories of macroscopic objects: inanimate matter, which can exist mainly in three different aggregate states—gas, liquid and solid, and biological organisms (animate matter), which differs significantly from inanimate matter in terms of its structural organization.

That description covers pretty much all that exists in the world at some fundamental level of understanding. There are only a few kinds of stable particles, namely, protons, neutrons and electrons, which make up all physical objects, and in addition there are neutrinos and photons, which are particles that mediate the interactions between the stable particles, and as we saw, can be considered epiphenomenal. There is also a huge number of unstable, short-lived particles, which however do not affect much the behavior of the stable particles and the evolution of material structures. Most of the particles aggregate in clusters and therefore interact strongly and frequently with each other, leading to rearrangements of the structures which constitute an evolutionary process of material structures that has produced ever more complicated structures in the form of living organisms, culminating with the most complex one—the human being—which is endowed with the capacity of reason and consciousness.

At that level, the description of the world is rather simple, and the vast majority of the conceptual framework that makes up our understanding of the world relates to regularities in terms of structures and processes involving large aggregates of particles, i.e., macroscopic objects. Besides the concepts denoting physical entities, our conceptual framework is considered to contain also abstract notions, such as love, peace, purity, similarity, etc. Most of those abstract concepts actually refer to properties or processes belonging to the realm of the human mind. Assuming that the human mind is fundamentally a physical entity (and we will see how consciousness can be explained in terms of physical phenomena in the next chapter), it means that the abstract concepts can also be reduced to physical matter, i.e., can be defined in terms of movements and interactions of particles (most of which would occur in the brain). So, it turns out that all of our understanding, our entire knowledge of the world, can be expressed in terms of aggregations of particles and their behavior! That is how we need to picture everything the world, and we will try to use this approach in the forthcoming discussion.

We saw also that particles can be understood as entangled knots of representations. Since all our knowledge can be reduced to the movement of particles, it means that it can also be reduced at an even more fundamental level to interactions of representations. Thus, the notion of representation turns out to be very useful in our understanding of the world and, as we will see later, since it can be construed both as a physical and as an abstract notion, it can serve as well for our understanding of consciousness and the realm of mental phenomena. In this way it becomes even more fundamental than the notion of matter and to some extent replaces it.

Particles are the simplest representational structures. Particles of the same kind share some common structure, and therefore behavior, akin to the way biological organisms from the same species share common structure and behavior. They, however, are aggregations of particles, and therefore their structure and behavior (broadly construed, i.e., any transformation of the structure of particles that forms the organism) are more complex. Nevertheless, due to the similarities of the structures and their transformational processes, there are common patterns in the structural organization and the behavior of the complex entities (both living and inanimate), which we can notice and about which we can form a concept in our mental knowledge framework. For example, we have a concept of eating, which is some kind of regularity in the behavior of certain aggregates of particles which we denote as animals. Inanimate objects have a smaller set of behaviors, but there are still many regularities that we can note, such as the movement of an object under the force of gravity, which we call falling, etc. There is an even larger (much, much larger) set of behaviors which we do not conceptualize, which are too idiosyncratic to form a very distinct pattern, and which we can conceptualize to some extent through a longer description involving multiple concepts. These are the individual events that involve inanimate and living objects, i.e., aggregations of particles, and which are unique in some respects, e.g., a particular instance of eating a carrot.

So, basically, our picture of the world should be one of particles *qua* representations, which interact and form aggregations due to their interactions, and in doing this give rise to regularities in their structures and behaviors which we can note and form a concept of. In fact, from the point of view of particles as representations the regularities are not just accidental—it seems natural for a system of interacting representations to produce regularities, since the very nature of a representative relation involves some similarity, and complex structures made up of representational relations would also necessarily possess some degree of similarity, i.e., shared structure and behavior. Thus, the repeated patterns in the structure and behavior of matter are a consequence of its very nature as interacting representations.

In the coming sections we will examine how the regularities give rise to life and eventually to the brain and consciousness. This development looks much more natural within the framework of representations than the classical picture of matter as tiny objects moving in empty space, which we considered in the beginning of the chapter. Our goal would be to replace the classical picture of matter with the representational picture developed thus far, which would allow us to answer the deeper questions posed in this book—what is consciousness?, what is the universe?, and, most generally, what is the world?

Aggregate States of Matter: Gas, Liquid and Solid

In the conditions typical for the surface of the Earth, which we said are of most interest to us, particles are relatively densely packed and therefore frequently interacting. That is to say, they have relatively high kinetic energies compared to, e.g., the particles in interstellar gas clouds. In the framework of our picture of matter as representations this can be stated as a relatively high degree of irreconciliation among the representational complexes making up the particles. The reason for this is the force of gravity, which makes the particles clump together forming the huge ball that we call planet Earth and consequently making their interactions very frequent.

The force of gravity in our picture of representations is a sort of memory of the representational complex for its earlier state, i.e., a tendency to retain part of its earlier configuration in the current configuration. When this happens for two particles engaged in an interaction, they both tend to retain the shared part of their states and thus tend to move towards each other in state space. However, when many particles move towards each other, they put more pressure on each others' states to get reconciled. Since they started with some degree of irreconciliation among them, and since representational relationships cannot be created out of nothing or disappear (according to the conservation laws), but can only be transformed, the closer packing of particles in state space means that they would need to affect each others' states to a higher degree. To state this in another way, a looser packing of particles in state space means that they all have a certain degree of differences in their states given by the distances between them in state space. When they come together in a denser packing under the force of gravity (i.e., as a by-product of their interactions), their states move closer in state space, and more specifically, the representational configurations of their self-reflecting cores become more similar. However, since the degree of difference must remain the same due to the conservation laws, the mutually reflecting representational components in the periphery must become more dissimilar. This means that they would put more pressure for change on each other and thus affect each others' states more strongly. In terms of the classical physical conceptual framework this is expressed as a higher kinetic energy, i.e., faster moving and more strongly colliding particles, and consequently higher temperature of the material structures.

This is basically what happens with the particles making up the Earth and any other cosmic body, like the Sun and other stars and the planets around them. The Earth has a higher temperature than the interstellar space surrounding it, and the temperature gets higher and higher as one gets closer to the center of the Earth.

The mean kinetic energy (which determines the macroscopic parameter of 'temperature') and the density of the packing of the particles (which produces the macroscopic parameter called 'pressure') determine the character of the movements and the interactions among the particles. When the particles have a lot of space to move around (low pressure) and a lot of kinetic energy (high temperature), they tend to move undisturbed for a relatively long period of time until they meet another particle, recoil in a brief encounter, and then move again freely. This kind of behavior represents the aggregate state of matter called 'gas.' When the particles are very densely packed and have low kinetic energy, they have very little space to move around and are also not affecting each others' states as strongly, so they tend to keep their configurations. This is the solid aggregate state of matter. Between those two aggregate states there is an intermediate range where the particles are relatively closely packed, but they have some freedom to move and rearrange, and this represents the liquid state of matter.

The basis for these three types of behavior of aggregations of particles is the U-shaped curve representing the energy of the system of two particles with opposite electromagnetic charges which are experiencing an attractive force. When two particles (any two fermions) are packed close together, they experience a repulsion force due to the fact that they try to retain their individualities, which precludes them from occupying exactly the same state and consequently exactly the same location in state space. At a somewhat longer distance, on the order of the radius of the particle, they experience an attractive force due to their electromagnetic charges. At this distance their states are different enough so they can easily retain their individualities, but similar enough so that they would be engaged in a relatively strong interaction (which could be described as exchanges of virtual particles). The interaction is attractive in the case of opposite electromagnetic charges and repulsive in the case of same charges, which reflects the symmetries in the configuration of the representations making up the particles. As the distance between the particles grows, their states become more and more dissimilar and they affect each other's states less and less (the exchanges of virtual particles become less frequent). Thus, the strongest interaction between them occurs when they are relatively close together, but well separated, which is the typical arrangement of particles in the solid and liquid aggregate states. In gases the movements of the particles are so fast that they fly through the relatively small regions of high attractive force very quickly and spend most of their time at large distances from each other, where they experience very little attractive or repulsive force.

The solid and liquid aggregate states are collectively called 'condensed matter' and are characterized by relatively frequent interactions among the particles, which can be regarded in fact as a continuous interaction process. This gives rise to a much richer set of phenomena related to the behavior of the particles and the complexes they form compared to that in gases. For example, regular arrangements of atoms in a crystal lattice may produce 'quasiparticles', i.e., behaviors which travel like a wave through the lattice and which resemble a real particle with novel properties, different from those of the actual particles making up the lattice. Such quasiparticles are a very natural phenomenon if we regard particles as representational complexes. In this framework, a regular arrangement of representational complexes would produce also a regular arrangement of symmetric relationships, and any disturbance of one of those relationships would cascade through the system. Since the moving disturbance is a representational relation in its own right, it can also be regarded as a particle and its properties (e.g., mass, spin, charge, etc.) can be measured and calculated as it is done for any of the conventional particles.

The quasiparticles are just one of the phenomena of interference of the wave functions of the frequently interacting particles in liquids and solids, which can also be understood as reconfigured representational complexes where the conceptual dividing line between two representational complexes that constitute different particles is drawn in an unusual place. In general, the continuously interacting representational complexes in the case of liquid and solid aggregate states of matter can be regarded as networks of representations, i.e., one giant and very complex particle, rather than consisting of a multitude of basic constituents (protons, neutrons and electrons) which interact in complex ways. It is true that the constituent representational complexes retain their identity, i.e., the self-representational core remains the same in the conditions typical for the surface of the Earth, but the peripheral representational elements regroup and aggregate in so many different ways that the boundaries between the original representational complexes become smeared and to some degree meaningless. When one of them segregates from the giant network it acquires again its typical properties of a particle in isolation, but when it is part of the network it changes its character and behavior according to the pressures from its environment.

This kind of intermingling of representational elements and the emergence of resonances, i.e., 'virtual' representational complexes, made up of elements of originally different representational complexes belonging to different basic particles, is possible only in the liquid and solid aggregate states in the conditions typical for the surface of the Earth, but not in the gaseous state. Besides those three aggregate states, in extreme conditions which are not found on the surface of the Earth, such as very high or very low temperature, particles' behavior changes and forms other aggregate states, namely, plasma in the case of very high temperature and Bose-Einstein condensate in the case of very low temperature. The plasma state consists of 'bare' charged particles which are not able to stick together and form atoms and molecules due to the very high kinetic energy associated with the high temperature. This aggregate state is typical for the matter making up the stars, including our Sun, and is in fact by far the most common aggregate state of all the matter in the Universe. It can be produced relatively easily also on Earth, but that requires special apparatus, i.e., special arrangement of physical structures, which in fact are aggregations of particles mostly in the solid state of matter.

The Bose-Einstein condensate can also be produced on Earth by use of special apparatus, and in fact does not occur naturally anywhere else in the Universe since it forms only at extremely low temperatures which are not found naturally even in outer space. However, it is a very interesting state of matter, since in this state all particles lose their identities and join in one large quantum complex behaving like a single particle. This is the ultimate case of representational elements mixing up and regrouping—they form one giant, neatly arranged network with very little movement of its elements, i.e., it seems like being frozen in time. This state of matter has also some very curious properties, such as delaying the speed of light travelling through this medium to such extremes that it can almost stop completely. This phenomenon has important practical implications since it can be used to perform quantum computations which can be much faster than their classical counterparts for some specific computational problems.

The two extreme states of matter will be of interest to us later on, when we consider the totality of all matter in the Universe and its future evolution, but for the moment we will focus on the material structures on the surface of the Earth for which we know that they are capable of producing life and ultimately consciousness.

Biological Matter (What is Life)

The surface of the Earth is characterized by temperatures (i.e., mean kinetic energy of the interacting particles) which are high enough to keep the materials (i.e., particle aggregations) naturally found on Earth in a solid, liquid or gaseous state, but not high enough to turn everything into plasma, as it is the case with matter making up the Sun and the other stars. Most importantly, the temperature on the surface of the Earth is just right to keep the molecules of water in a liquid state—preventing them from sticking together in a crystal form (ice) of flying apart as a gas (vapor). Since water is abundant on the surface of the Earth, and Earth is of the right size to have an atmosphere which allows for water vapor to circulate and create the phenomenon of rain, the conditions are right for the constant mixing of materials in the solid and in the liquid aggregate states aided by the actions of atmospheric gases. All this mixing seems to be a key requirement for the emergence and evolution of life forms.

From the point of view of particles as representations, we stated that the action of gravity is to bring initially dispersed particles closer together, which increases their temperature due to the fact that the degree of dissimilarity of the representations remains the same while they move closer together in state space and thus the degree of similarity increases. In classical physical terms, this process is understood as an increase in temperature due to friction, i.e., more frequent and more violent interactions between the particles. So the increased frequency and energy of the interactions among particles translate into larger pressures on the representational elements from the periphery of the representational complexes of the particles, stemming from the conservation of the degree of dissimilarity as the particles move closer together.

When the particles lock in a stable configuration in the solid aggregate state of matter, their representational complexes effectively fix their environments, keeping the representational interactions with their neighbors and rarely engaging in interactions with particles further away. However, due the relatively high temperature, the representational interactions are constantly transforming in a chaotic fashion, which looks like vibration of the particles if we try to observe their positions by bombarding them with photons or electrons. So, the original degree of dissimilarity from the initial configuration of matter in the universe translates into high temperature and constant chaotic transformation of the representational interactions in the solid aggregate state. In this state, the particles have a fixed representational environment, but their states constantly loop through a confined region in state space without being able to form reconciled representational complexes.

The opposite is true for the particles in the liquid state of matter. Their interactions are too weak to lock them in stable configurations, so they constantly rearrange their positions and break up and recombine their representational connections. This state too preserves well the original degree of dissimilarity, since the movements and the interactions among the particles are also chaotic, meaning that each act of reconfiguration of a representational complex due to pressures from the environment is unrelated to the other acts, keeping it irreconciled with them and thus preserving the total amount of irreconciliation among the representations.

The only way for the representational complexes to get more reconciled (which we posited to be the main tendency in the evolutionary process of all matter in the universe) in the conditions of high temperature prevalent on the surface of the Earth is to lock their transformations in a coordinated synchronous process. In this state, instead of switching constantly from one representational connection to another via exchange of photons, the particles' states can evolve in parallel, undergoing similar transformations of the state at least for some limited period of time. This parallel evolution of their states gives rise to repeatable processes involving the same structures in every run of the process, and if the process starts with one set of structures and transforms them to another set of structures (i.e., molecules, or molecular complexes), then the process effectively results in copying structures starting from the same basic ingredients.

In short, the representational nature of matter which always strives to achieve a greater reconciliation of the representations leads to processes of reproduction of structures and of the processes themselves, which is arguably the main hallmark of living systems. It is more problematic to imagine how this works if we are using the classic picture of matter as tiny spherical particles moving and colliding in empty space and sticking together under the attractive electromagnetic forces, and the necessity of the emergence of reproductive types of processes appears much more natural within the framework of particles as representations. In this case the key requirement for the emergence of reproduction and life itself is a large variety of structures and processes situated in relatively stable physical conditions (in terms of average energy of the particles, i.e., temperature, low radiation, etc., and average density of the particles, i.e., pressure due to the strength of the gravitational force). The variety of structures and processes allows for many possible arrangements to be tried in the course of the evolutionary process of matter, and the stable conditions allow for the persistence of the successful structures by limiting the range of possible states that matter can occupy.

We can call the parallel evolution of the states of particles or structures of particles an *adaptation*, which can relate both to the structure of the complexes or the processes in which they are involved. This differentiation of structural and temporal adaptations is not dichotomous, and is for the most part an abstraction which only serves to aid our analysis by simplifying the real picture of the natural events. In reality, the representations adapt to each other both structurally (in space) and as processes (in time) at the same time, and the distinction between the two types of adaptations is arbitrary.

We can also call the adapted structures *representations* in their own right, since they constitute perpetuated representational relationships. They are more complex than the basic representational elements making up the fundamental particles, and since there is a great variety of possible arrangements of structures made up of the basic particles, there is also a great variety of possible complex representations. Each of those complex representations can be considered a particle, just as any of the basic particles is an intertwined network of representational connections, but unlike the basic particles, the complex representations cannot exist in isolation, i.e., they are dependent on the basic particles for their existence and if we try to isolate them, the networks of representational connections would reconfigure again into the individual structures of the basic particles. Therefore, the complex representations are epiphenomena supervenient (i.e., depending for their existence) on the basic particles.

The reproduction of processes and structures requires frequent rearrangement of structures of condensed matter, but on the other hand it requires also some degree of persistence of the structures. In this way, the organization of living matter needs to bear partly the characteristics of both the solid and the liquid aggregate states of matter. Indeed, virtually all life forms are structurally in between the solid and the liquid state—they have persistent overall shape which is slowly transforming in the process of aging of the organism, starting from a small structure (a germ, embryo, egg, etc.), growing to a mature organism and then declining and eventually disintegrating after the death of the organism. All organisms contain water, which is a liquid, as part of their structures, but the water molecules are bound to other molecules in compounds, so the organism as a whole has a persistent (but transforming) structure, which sets it apart from the typical liquids.

We can understand the process of transformation of the organisms from birth to death, which in fact encompasses completely our concept of 'life', also in terms of the transformation of structures of complex representations. The initial process of growth from inception to maturity is a process of increasing coherence of representational structures in which the organism takes in unadapted structures from the environment (through food, breathing, osmosis, etc.) and converts them to highly adapted structures incorporated as part of the organism itself. The growth of the organism, however, as a complex of adapted structures, results in a gradual process of decoherence due to the constantly accumulating diversity of the structures and processes (we can call them also representations) interacting with and being taken in by the organism. Thus, the constant impact of diverse representations from the environment gradually erodes the coherence of the representations making up the organism and leads to its death and disintegration.

In very general, abstract terms, we can say that life forms represent concentration points for the initial degree of similarity of the representational structures making up the universe, due to the fact that the representational structures stay more reconciled when they are locked in the coherent transformations that constitute the life forms, but the rapid and forceful reconciliation of representational structures in the initial stages of the life of an organism draws in also a high number of unadapted representational structures, which reflect the initial degree of difference of the representational structures in the universe, and this leads to the dissipation of the highly concentrated representational structures, culminating in the death of the organism. On average, though, the life process is a process of gradual accumulation of representations, and the cycle of life and death of the organisms is just an oscillating, pulsating deviation from the slowly advancing average.

When talking about biological structures of particles and processes we need to bear in mind the difference in the temporal and spatial scale compared to the structures and processes pertinent to single particles which we discussed so far. This difference spans multiple orders of magnitude, from the miniscule scales of the quantum physical phenomena to the macroscopic scale of our experience. Although many of the biochemical reactions inside the organisms of living beings happen extremely fast, on the time scale of the interactions between individual particles, they need to affect a high numbers of particles in order to alter significantly the behavior of the organism, so their effects become visible mainly on the time scale of human experience. For example, the macroscopic events of contracting a muscle and moving a limb take hundreds of milliseconds, while the microscopic events of the transfer of an electron from one atom to another, which underlie most chemical and biochemical reactions, take about 1/3 of a femtosecond (10^{-15} s). This is a difference of about 14 orders of magnitude, and the differences in the spatial scale between the size of a muscle and the size of an atom are similar—about 9 orders of magnitude.

These large discrepancies of scale must be kept in mind when we imagine the structures and processes at work in biological and inorganic matter. This means that we should imagine the familiar macroscopic scale phenomena as composed of very large numbers of simple interactions between individual particles, and what we observe through our senses actually are mass movements and actions unfolding very slowly from the perspective of individual particles. This provides a lot of room for a large variety of structures and processes involving aggregates of particles to exist in the intermediate scales between the macroscopic and the single particle extremes. Without the appropriate understanding of the scales, our imagination of the physical processes and structures would be inaccurate and it would be hard to comprehend what makes this huge variety possible.

Evolution as Accumulation of Representations

When we examined the behavior of the representations that constitute all particles of matter, we concluded that they strive to achieve more reconciled states and this is the main characteristic of the time evolution of the state of all matter. This idea translates also to the larger scale of aggregations of particles and complex representations. Indeed, we saw that biological structures themselves are a result of the pressure for a greater degree of reconciliation which finds an outlet in coordinated transformations of particles' states. So, the biological organization of matter can be regarded as a higher concentration of reconciled representations compared to a mass of inorganic matter with the same composition. In other words, the representational interactions in the biological structures are more coherent and constitute complex representations to a higher degree than their inorganic counterparts. Thus, we can say that the biological structures have a higher degree of representational relations, and since every macroscopic biological structure grows from a simple, small unit (germ, cell, egg, etc.), it means that the representational relations must accumulate in the process of growth of the organism.

The growth of an organism is a reflection of the evolutionary transformation the species has undergone in the long process of evolution in biological terms. Thus, the species themselves, and the totality of all life forms as a whole, may be considered as undergoing a transformation which at the most fundamental level can be understood as a process of accumulation of representations. Let us consider how this works in greater detail.

According to modern evolutionary theory, life originated on Earth about 3.8 billion years ago, and the simplest known life forms were prokaryote cells resembling present-day bacteria. These cells have also the simplest mechanisms for functioning and reproduction. Their genetic material is dispersed throughout the cell, instead of being organized into chromosomes in a cell nucleus, and they grow by intake of organic substances and divide when they reach a critical size. Sometimes they merge and exchange some of the genetic material in a more or less random fashion.

These unicellular organisms are asexual and practically eternal, since after the cell division it is not possible to specify one cell as the parent and the other as the offspring. They multiply in the presence of food and die only due to starvation, chemical toxins or mechanical damage, but not due to aging. For that reason, we can regard them as non-aging replication mechanisms, unlike the multicellular organisms which undergo a cycle of birth and death.

In terms of our picture of life as concentration of complex representations, we can regard these unicellular organisms as the simplest form of organization of particle structures which has achieved a high enough concentration of representations in order to replicate itself. To state this more precisely, as the particles structures become more adapted to each other, the transformational processes that they undergo in a coordinated fashion become more in number, longer in duration (i.e., longer chains of biochemical reactions), and more interconnected. All this leads to the formation of a more complex structure which works as a single mechanism. In terms of representations, we can say that diverse representations aggregate because of the action of their similarity components, thus increasing their total degree of similarity.

On the other hand, as the complex gets bigger and bigger, the dissimilarity components aggregate as well, and they put more and more pressure on the complex representation to dissipate. Thus, we have two opposing tendencies—the aggregating adaptations based on the similarity component of the representations tend to unify more and more diverse structures, while the difference component of the representations expressed in the diversity of the structures and their behavior tends to disrupt the coordinated transformational processes.

These two pressures result in the eventual splitting of the representational complex in two in the process of cell division when it reaches some critical mass. Also, since the coordinated transformational processes tend to have the same inputs and outputs in terms of particle structures, they effectively work as replication mechanisms for particle structures. In this way, there is an accumulation of redundant structures adapted to the same complex representation (i.e., the mechanism of a single cell), so eventually there is enough building material for two complexes of adapted representations, and this leads to the splitting of the cell into two copies of itself. This is the basis of the reproductive function of the cell.

In more abstract terms, we can say that the overall degree of similarity and difference of the representational nature of matter is for the most part conserved in this process, with a slight increase of the degree of reconciliation of the representations. When the cell grows, the similarity component grows disproportionately larger than the difference components, which remain hidden in the individual interactions among the particles making up the biological complex, but when the cell divides, the difference degree is restored at the larger scale of the complex representation that constitutes a cell, since there are two individual structures which are highly adapted internally, but only very weakly adapted to each other. On the other hand, the similarity component grows a bit more, due to the similarity of the structures of the two complex representations constituting the new cells, and the resulting possibility for interactions and exchange of material between them.

The next step in the evolutionary process is the increased complexity of the mechanisms of the cells, resulting in more genetic material which gets packed compactly in a nucleus, the incorporation of foreign structures into the cell, such as the mitochondria, and the formation of substructures (organelles) with their own functions. Again, this process can be regarded as an expression of the two opposing tendencies of accumulation of similarities and differences. As the size of the mechanism made up of adapted structures and processes grows, the variations in the degree of adaptation between particle structures within the complex grow as well, leading to the specialization of function and diversification of the adapted structures. Some particle structures become more adapted to each other, i.e., the complex representations they constitute become more concentrated, and they form functional units which are weakly adapted to the mechanism as a whole. Thus, similarity grows within these units, and the degree of difference is expressed by the functional and structural specialization of these units.

If we imagine the complex representations as inhabiting an imaginary state space made up of all possible states of the organisms, then we can imagine this process as a sort of gravitational clumping of the particles (representing the complex representations) and thinning of the density of the particles at the boundaries of the concentration regions. This reflects the functional specialization of the physical structures within the cell—functionally they become more tightly coupled within the functional unit, but more dissimilar to the structures of any other unit, yet still functionally coupled to them. Also, we can understand the replication phenomenon related to physical structures as a kind of self-representation in the abstract state space. Indeed, two (nearly) identical physical structures would occupy the same location in the abstract state space, and if they interact, that would look like the oscillation of a self-representation in the abstract state space.

The next big step in the evolutionary process is the emergence of multicellular organisms. From the point of view of representations, this occurs when there is enough variety and concentration of cellular structures generated by the constant accumulation of adaptations, so that the diverse cells and their life processes are adapted to each other enough that they can start functioning as a unit.

The emergence of the multicellular organism is a complex process based on the basic process of division of the cell. For that reason, the representational units that accumulate are the temporal representations of the division process, rather than the spatial representations of particle structures adapted to each other, which is the case in the formation of a cell. That leads to a different basic organization of the evolutionary process of accumulation of representations. The direction of the accumulation of adaptations shifts from the lateral expansion of biological structures which convert inorganic structures into biomass by replicating themselves (cell division) to the vertical expansion of replicating processes of growth of multicellular organisms.

The replication of a process allows for faster transformation of the evolving structures and also produces a greater variety of structures as a by-product. This is due to the fact that slight modifications to the process still result in highly adapted, and therefore biologically viable, structures, while the modifications in the laterally replicating structures are less likely to be adapted to each other and to be viable.

In this way, the replication of the process of growth of the multicellular organisms results in faster evolution and faster accumulation of representations, but it also results in the temporary existence of the biological structures—the cycle of life and death of organisms. Since we have a replication of a process of growth, this process must constitute a unit, i.e., it must have bounds. In the case of a process, the bounds are temporal, i.e., it has a beginning and an end, therefore, the biological structures formed by this process that constitute the organisms also have a beginning and an end. This arrangement, however, allows for the faster accumulation of representations, meaning increasing degrees of both representational similarity and difference in the life course of the organism, with a slightly higher advantage for the similarity component compared to lateral replication of structures.

The final milestone in the advancement of the evolutionary process is the emergence of the brain and the conscious organization of matter. As we will see later on, the phenomenon of consciousness is based on an even higher degree of representational similarity at the microscopic scale of particle structures. This increased degree of similarity needs stable conditions, since it needs to counteract the chaotic processes in the relatively high-temperature conditions on the surface of the Earth. The evolutionary development of the maintenance of fairly constant temperature of the mammalian (and in particular, the human) body provides this stability and enables the even faster evolutionary process of consciously organized matter.

Within the framework of representations comprising of components of similarity and difference, the classical ideas of evolutionary theory of mutations and natural selection can be reduced also to variations in the degrees of similarity and difference among particle structures. The organisms are particle structures with high concentrations of representations, and therefore large-scale similarity of complex representational structures. The degree of difference hidden in the fine-scale structures constantly perturbs the replication of the growth process of the multicellular organisms and results in mutations. The biological organisms are also to some degree adapted to the environment, consisting mostly of inorganic matter being constantly reorganized by natural processes fuelled by the energy from the Sun. The constant variations in the natural conditions in the environment, as well as the interactions with biologically incompatible organisms (predators, poisonous prey, etc.), are an expression of the degree of difference in particle structures, which can concentrate and result in radically different conditions that can destroy an organism. This phenomenon is captured in the notion of natural selection, although violent death of the organisms is relatively rare and most of the pressure for evolutionary change comes from the persistent interactions of maladapted structures, e.g., variations in climate and food resources in different geographic locations.

In conclusion, the foregoing discussion suggests the assertion that the evolutionary process can be completely understood only within the conceptual framework of evolution as accumulation of representations, and the classic conceptual apparatus of evolutionary theory, although very useful for practical purposes, is not adequate for a complete explanation of the workings of the evolutionary process and the reason for its existence. It is also inadequate for justifying the emergence of consciousness sustained by the special kind of organization of particle structures in the brain, which will be the topic of discussion in the next section and the entire next chapter.

The Brain as the Culmination of the Evolutionary Process

The brain and the phenomenon of consciousness seem to be a product of the evolutionary process. They appear at some level of complexity of the organisms as the evolutionary process accumulates more and more adaptations resulting in more and more complex structures. They also evolve from simpler to more advanced stages, with human consciousness being the most advanced form of consciousness compared to its animal counterparts.

So, we need to think of consciousness as a kind of phase transition occurring with the accumulation of adaptations, or, in other words, with increasing concentration of representations. This line of thought suggests that the nature of consciousness ought to be related in some way to representations, which we concluded are a good conceptual tool for understanding the nature of matter. We imagined representations and their interactions forming a kind of abstract state space which behaves just like real physical space as we know it from the point of view of modern physics, and we can use the same "trick" to imagine what happens with the complex representations that concentrate in the life forms and ultimately produce consciousness.

The picture of this second-order state space, which is made up of states of particle structures in physical space, is in some ways similar to that of physical space. The complex representations in inorganic matter would constitute separate, isolated states which interact from time to time, much like the particles in a gas. So, inorganic matter (even the condensed one making up the stars and the planets) would correspond to a gas in the second-order state space. The life forms, on the other hand, represent increasing concentrations of such particles, much like the clumping of the gas in the universe into cosmic bodies—stars and planets. Consciousness, then, is like a phase transition in this clumping process, where it starts to interact with the structures in its environment, much like consciousness manifested through human and animal activity modifies its physical environment and results in rearrangement of particle structures.

This line of thought will be pursued further in Chapter 3, after we have gained some more detailed understanding of the workings of the brain and the phenomenon of consciousness, which are the topics of the following chapter.

CHAPTER 2: WHAT IS THE BRAIN

Basic facts about the brain

The brain is one of the organs in the human body and it is structurally more complex compared to most other organs, like the liver, muscles, lungs, etc. It bears characteristics of both the solid and the liquid aggregate states of matter, with an ability of reorganizing some of its structures.

The brain, as all organs in the human body, consists of cells, but in the brain they are special kinds of cells unique for this organ. The most important kind of cells are the neurons, which come in several shapes and sizes, but the common feature of all of them is the fact that they have protrusions, called axons, which propagate an electric signal via a rapid change in voltage across the cell membrane travelling like a wave down the axon. This phenomenon is called neural firing and it serves the purpose of influencing other neurons' internal states.

The axons most often make connections with another kind of protrusions coming out of neighboring neurons—the dendrites—forming synapses at the touching points between them. In this way, the electric voltage signal travelling down the axon is transformed into a chemical signal at the axon terminal through the release of a neurotransmitter (special kinds of small molecules which have the ability to bind to other, larger molecules in the dendrite terminal and thus affect ion concentrations and therefore the voltage inside the other neuron), and then to a voltage signal again in the dendrites and the body of the neighboring neuron.

This basic mechanism allows the neurons to form very large and complex networks where each neuron influences the state of hundreds of others and is in turn influenced by hundreds of other neurons. The neurons typically connect to their neighbors, but there is also a very well defined (and very complex) pattern of long-range connections between neurons from different brain areas.

Most of the important neurons—those involved in conscious thought—reside in the upper layer of the brain, called the cortex, which is a sheet of cells, about 2mm thick, at the surface of the convoluted brain tissue. The inner tissue beneath the cortex is made up mostly of axons projecting from neurons in one brain area to another brain area. Dispersed among the neurons and the axons are glia cells, which serve mostly the purpose of maintaining the structural integrity of the brain, but also act as storage of neurotransmitters and influence the neural states in other ways, different from the firing mechanism.

As all other cells in the human body, the neurons are produced by division from the initial zygotic cells. They acquire their mature characteristics and their physical location within the brain after a process of differentiation, during which different cells take on different evolutionary paths of development. Still, what is important about the ontogeny of neurons and, indeed, all the cells in the body, is that they are created by a process of cell division involving replication of molecular structures, and therefore they are entities with similar composition and functional relationships between their constituents (macro—and micromolecules).

All the cells in the body cooperate and complement each other's functions, creating a mutually supportive environment in which they can survive and multiply. Besides that, neurons also affect each other's internal environment and functioning through the mechanism of neural firing described above. This mechanism is unique for the neurons and the cells in the rest of the body have no comparable way of affecting each other's state with similar efficiency. They can only signal to each other through the release of chemical substances, but this is a much less efficient and slower way of communicating.

The brain is also special because of the fact that it needs relatively more stable conditions to function properly. It can tolerate smaller deviations in temperature and oxygen supply than the other organs, and for that reason the body has a number of physiological mechanisms to ensure those stable conditions. Any significant deviation from the homeostasis results in loss of consciousness and soon after that in mass cell death in the brain tissue.

In general, for the purpose of understanding what consciousness is and how it came about, it is important to keep in mind the evolutionary history of the brain as one of the organs of biological organisms belonging to the animal kingdom. This evolutionary history needs to be understood in terms of the process of aggregation of representations, which, as we saw in the previous chapter, underlies the evolutionary development of all life forms in general, and in the case of the brain leads to the highest concentration of representations (in terms of adapted particle structures, which themselves should be considered complex representations). The special mechanisms of neural firing and the regulation of blood flow and temperature should also be conceived within that framework—as mechanisms ensuring the effective interactions in a physical system with a high density of representations.

What is Consciousness

Consciousness as a notion is typically related to the first-person experience of perceiving the world and one's own stream of thought. It is the character of that experience—the vividness of the visual and auditory sensations, as well as the self-awareness of one's own thoughts and emotions—that baffles many people and poses a seemingly insurmountable explanatory challenge to them. There is nothing else like it in the world—they would argue—and all our explanations of physical phenomena in the world are inadequate for the task of explaining what is consciousness and why it exists.

That belief, however, is based on the classical notion of matter that we presented at the outset of Chapter 1. If we imagine matter as representations, on the contrary, then we will see that consciousness not only may be adequately explained, but in fact is a natural consequence of the representational nature of matter and plays a very important role in the ontological description of the world. That role will be explored in more detail in Chapter 3, and for now we will focus on building the conceptual framework for imagining the conscious mental states as physical phenomena using our familiar idea of the representational nature of matter.

In order to understand what consciousness is, we need to consider two basic ideas that we established in the previous chapter. The first one is the representational nature of matter, i.e., we need to imagine particles not as tiny balls flying in empty space, but as complexes of representations which form the epiphenomena of space and time through their interactions, entailing phenomena like entanglement which represent non-local shared states between spatially disjoined particles. The second idea is the understanding of life forms as structures with increased density of representations (in the form of spatial and temporal adaptations), and of the brain as the culmination of this process.

That view suggests that the phenomenon of consciousness appears as a consequence of a phase transition in the process of accumulation of representations when they reach some critical density. That means that we need to look for the explanation of the consciousness phenomenon at the level of individual particles and their interactions in the conditions of highly adapted (both spatially and temporally) particle structures.

So, what happens to particles when they form highly adapted structures and how should we imagine such structures in terms of our picture of particles as representational complexes? We saw in Chapter 1 that in the picture of particles as representational complexes the representations are in constant flux—they form and dissolve between different complexes in quick succession, while they transform but remain interconnected within the complexes. The hallmark of life was found to be the expansion of the representational connections both spatially (as adapted to each other structures) and temporally (as adapted to each other processes) in aggregations of particles.

In inorganic matter, therefore, no matter how large is the aggregation of particles, the representational connections are short-lived and disjoint, i.e., there is no continuity between two successive connections. In organic matter, on the other hand, there is some degree of continuity, lasting during the time the adapted structures or processes cohere, i.e., while the representational relations between complexes remain interconnected. This is made possible by the greater degree of adaptation, which by definition means that the complexes represent each other to a greater degree and therefore the representational relations between them contain a higher degree of similarity and consequently are more robust and longer-lasting.

Now, as the diversity of particle structures increases, the degree of dissimilarity among them also increases in the general case, but if that happens in the course of evolution to structures making up the bodies of living organisms, then there is an opposite tendency for these structures to become more adapted, resulting in a greater variety of highly adapted structures and processes. In humans the greatest variety of molecular structures is found in the brain, as indicated by the fact that the highest number of genes in the human genome encode proteins expressed in the brain. Thus, the brain is the organ in the human body with the highest concentration of variety of adapted structures. We can think of the brain also as the structure with highest concentration of representations, since we define a representation as a persistent representational relation between two (in the case of simple representations) or more (in the case of complex representations) representational complexes.

So, we see that the particle structures in the brain are able to maintain more complex and longer-lasting representational relations among them. These relations get constantly created and constantly destroyed, producing a constant stream of coordinated transformations of particle structures, classically described as cascades of biochemical reactions. In simpler, i.e., evolutionary more primitive, brains the density of the representations is lower and therefore the continuity of the stream of coordinated transformations is less pronounced. It is punctuated by sharp transitions to new representational steady states, corresponding to a lower level of consciousness as we observe it in evolutionary more ancient animals. As the density of representations gets higher, the coordinated transformations get more complex and more coherent, producing the more advanced consciousness typical for humans, with more continuity in the transitions between the coordinated transformational processes.

Thus, consciousness is the persistence of representational relations as they are constantly transforming in the conditions of highly adapted particle structures at the relatively high temperatures typical for animal bodies. The key idea here is the continuity of the transformational process. Representational relations exist even between elementary particles, but they last on the order of femtoseconds and then they break apart and re-form again. We can consider this as some kind of a prototype of higher-order consciousness, i.e., consciousness at its simplest level. The representational relations in animal brains last much longer—on the order of milliseconds in insect brains, for example, and they are longest-lived in the human brain—on the order of hundreds of milliseconds, which is the typical time frame for the instantiation of an individual concept (such as any of the words in reading a text like this one).

The persistence of representational relations also entails the hypothesis that the particles involved in those representational relations possess shared states. In terms of our picture of particles as representations, this means that the representational relations connecting a particle with other particles in its environment transform without breaking apart for prolonged periods of time. In this way, a particle is continuously in a shared state with some other particles due to the persistence of at least some of the representational relations, forming a network of shared states with macroscopic dimensions. This network is a dynamic entity and particles constantly join it and leave it. Their involvement in the network is also graded and not dichotomous—particles may share a state with other particles to a different degree, and this determines how much they influence and are influenced by the shared state.

The shared state of two particles is akin to the phenomenon of entanglement, but also somewhat different from it. The entanglement of particle states typically refers to measurements of observable states which can be exploited for obtaining knowledge about specific properties of the particles at specific moments in time (e.g., whether an electron is spin-up or spin-down). In the brain, the particles do not need to be constrained to acquire definite observable states (i.e., the wavefunction does not need to collapse), so they can stay in an entangled shared state for prolonged (maybe even macroscopic) periods of time[7].

It is important to note that the shared states do not need (and probably do not) stretch over macroscopic distances and can only involve the immediate environment of a particle. However, due to the self-similarity of particle structures in the brain (i.e., the similarity in the molecular structures making up the cells and the processes going on inside the cells), there are many instances of particle structures undergoing the same (or very similar) transformation dispersed throughout the brain and therefore having a shared state with nearly the same configuration in multiple locations. These multiple instances of the same shared state are not isolated, but they influence each other through their interactions with the rest of the particles structures, which serve as a classical, non-entangled

---

[7] The longest coherence observed in biomolecules at room temperature lasts about 800 femtoseconds in a chlorophyl complex. That is a long way from the hundreds of milliseconds duration of a mental state, but the brain structures are much more adapted to each other than those in a photosynthesis molecule, and also the 800 femtoseconds duration is several times longer than what would be expected of an unadapted structure with the same physical parameters, which suggests that the phenomenon of 'dynamic adaptation' is real and just needs to scale up several orders of magnitude to produce conscious-like effects.

communication channel for relaying transformational pressures. This is possible because these other structures are also highly self-similar and behave in a coherent way.

So, our hypothesis is, generally speaking, that the high concentration of self-similar and adapted to each other particle structures and processes in the brain produces a very high number of instances of local entanglement of particle states which couple with instances of the same kind at long ranges through classical particle interactions that do not involve entanglement of their states. In this way, the multiple instances of entangled shared states in effect form one large shared state involving particle structures that can reside in any part of the brain. This is an entangled state with macroscopic dimensions, which however does not need a mechanism for maintaining continuous entanglement between individual particles at macroscopic distances.

The self-similar particle structures that realize the shared entangled state, which we denote as consciousness in abstract, non-physical terms, probably reside in neurons (and/or glia) of the same type, and therefore in particular brain structures. The empirical evidence indicates that the main locus of consciousness in the brain is the cerebral cortex—the thin layer of cells enveloping the outer surface of the brain. So, we can hypothesize that particle structures in this layer, which in its turn can be subdivided roughly into seven sublayers, engage in shared state relations and form a single virtual coherent shared state which is the physical substrate of conscious experience.

The neurons, as we said earlier, have the peculiar ability to communicate via the neural firing mechanism. This phenomenon most likely serves the purpose of modulating the shared entangled states of particle structures inside the neurons, allowing them to influence the unified coherent state more effectively, but it is probably not directly involved in instantiating consciousness. Synchronous firing seems like a suitable mechanism for locking dispersed particle structures from different cells in a shared state and in this way it may indeed represent a correlate of conscious activity. The levels of neurotransmitters in the synaptic gap junctions also play a modulating role by making it easier or more difficult for a neuron to trigger an action potential, which affects the synchronicity of the firing and the characteristics of the firing mechanism in general.

Another important factor for the maintenance of a complex conscious state is the stable temperature of the brain. The temperature is a global factor which affects the rates of biochemical reactions, the oscillation modes of molecules, the coupling strength of chemical bonds, etc., so keeping it as constant as possible allows the particle structures to retain their character and functions and therefore to maintain the continuity of the conscious state without switching from one mode to another (which effectively happens under the influence of intoxicating substances in the brain and also at elevated temperatures causing fever and delirium). Animals with stable body temperature have evolutionary advantage because of the continuity of their conscious state (i.e., they are more "intelligent" relative to animals with varying body temperature and therefore wildly transforming conscious state), and it is not surprising that one of those species—*homo sapiens*—developed the most advanced and most complex conscious state.

It is important to note that our picture of the representational nature of matter developed in the previous chapter is crucial for understanding what is consciousness and for imagining the shared entangled state. The classical picture of matter is largely inadequate for this purpose and has led many people to the deep conviction that consciousness is an irresolvable mystery. On the contrary, the representational picture not only explains consciousness in physical terms, but also offers a clue to the nature of qualia and phenomenal experience, which are particularly hard problems for the classical approach.

Indeed, if we imagine particles as tiny solid spheres flying in empty space, then it would be very difficult to imagine how their collisions give rise to phenomenal experience. This picture, however, is grounded in our experience of the macroscopic reality and as we saw is inadequate for imagining how things work at the microscopic level of individual particles. At this level, the particles and their interactions are best thought of as representations which stand in bilateral relations with each other and affect each other, giving rise to the epiphenomena of space and time. In an analogous way we can imagine them giving rise to the epiphenomenon of phenomenal experience, i.e., we can suppose the existence of some primordial, rudimentary property of representational relations which we refer to as phenomenal experience, and we can try to ascertain its characteristics and see if they fit our intuition of what phenomenal experience should be.

According to our intuition, phenomenal experience has a vivid experiential component which cannot be expressed with words, like the actual experience of seeing red or smelling a rose. These different bits of experience are called qualia. What has been established by modern science is that these experiences are correlated with physical processes that reside entirely within the brain, i.e., they are not a relation between a process inside the brain and the external reality, but rather a relation between processes contained within the brain, which in some way are related to, i.e., represent, external reality. Therefore, the qualia of phenomenal experience do not need to involve interactions between matter inside the brain and matter in the environment, but can be based only on local interactions, just like those between particles engaged in a shared entangled state, as we described them earlier.

So, where do the vividness of the experience and its complexity come from? The phenomenal character of the experience can be regarded as arising from the representational relation itself—when a material substance is representing other material substance it has a shared component with it, but at the same time it also has some differences with it. The shared component underlies the very capability for phenomenal experience, while the difference component shapes the character of that experience, i.e., the details which distinguish the sensation of seeing red from that of smelling a rose.

In this way, the character of the phenomenal experience depends on the complexity of the representational state formed by the interacting representations. The simple bilateral relations of individual particles engaged in an interaction, which we consider proto-conscious experience, cannot have the rich detail of the complex relations between representational complexes involved in a shared entangled state. Also, the shared entangled states can vary in complexity, which corresponds to our intuition that conscious states and conscious abilities vary in different animal species and in humans. The most complex, and therefore the most vivid and intellectually profound conscious states, occur only in the human brain because it sustains the most complex shared entangled states, while the other animal brains sustain more or less simpler states.

All these considerations suggest that the representational nature of matter itself should be equated with phenomenal experience and the various qualia can be mapped onto different configurations of shared entangled states of particle complexes. In the following sections we will develop this idea in more detail and we will try to learn to imagine the familiar aspects of mental life in terms of properties of the shared entangled state of particle structures in the brain.

How to Imagine Mental States

We said that the mental state is realized in the human brain by particle structures engaged in a shared entangled state made up of multiple instances of such particle structures distributed throughout the cortex. Since these are multiple instances of representations that reflect each other and form a unified state, we can use the same approach as we did regarding particles of matter and imagine this state as an entity situated in an abstract state space. If we do this, we will find that the mental state bears some similarity with the representations that constitute particles in three-dimensional space, and the main difference between the two is the level of complexity—the mental state is much more complex than the particle states, with many more degrees of freedom of transformation, rendering it effectively a representational structure situated in a very high-dimensional state space (on the order of tens of thousands of dimensions—as many as there are words and phrases representing individual concepts in a given language).

The fact that the mental state, imagined as an entity in abstract state space, bears the characteristics of particle states in physical space is not surprising given our hypothesis regarding the physical nature of the mental state, namely, that it is realized by particle structures with a shared physical state. Therefore, we can imagine the mental state very much like a particle with a high-dimensional complex structure, pulsating and transforming in state space. Its evolution through time looks very much like the evolution of the quantum state of a particle—going through successive phases of well-defined and more disordered states, corresponding to the phases of collapsing and spreading-out of the wave functions of quantum states of individual particles[8]. In contrast to the

---

[8] We can remind ourselves here of the observation Niels Bohr made regarding this analogy (already mentioned in footnote 3 on page 59): "…the apparent contrast between the continuous outward flow of associative thinking and the preservation of the unity of personality exhibits a suggestive analogy with the relation between the wave description of the motions of material particles, governed by the superposition principle, and their indestructible individuality. The unavoidable influence on atomic phenomena caused by observing them here corresponds to the well-known change of the tinge of the psychological experiences which accompanies any direction of the

particle states, the successive phases in the evolution of the mental state have macroscopic proportions—they engage large portions of the cortex and last hundreds of milliseconds, as already mentioned earlier. These alternating phases are mirrored roughly in the flow of language. When speaking or reading a text our mental state jumps from one well-defined state to another, corresponding to each word or phrase in the text, with brief periods of unrealized, less conscious transitional states in between.

We refer to the well-defined states as concepts, and each concept corresponds to a distinct type of particle from the second-order quantum realm of mental states. Just like particles are all representations in nature but with different structure, so the concepts are also complex representations in nature, but with different internal configurations. The mental state is continuously transforming, switching from one concept to another, but the concepts themselves stay in more or less stable relations to each other, forming more permanent configurations in the abstract state space. Thus, we can imagine the mental state of an individual person as traversing the conceptual structure in the abstract state space, but the totality of the mental states of all individuals is stable and determines the configuration and the evolutionary transformation of the structure.

In other words, the physics and the distribution of particles of matter in the higher-dimensional universe formed by the conceptual structure and the abstract state space associated with it are a result of the superposition of the totality of all mental states, with a single particle, representing a single concept, being a superposition of all possible (but not necessarily currently realized) mental states corresponding to this concept.

In this way there is a mapping from the states of particle structures in real physical space to smaller in size particle structures in the abstract state space of concepts. The transformational dynamics of the real particle structures in the brain also map onto the transformational dynamics of the mental states and the conceptual structures made up from them. These dynamics correspond to the train of thought, which we can imagine as made up of two components—one which comes into being (the front end of the mental state) and one which dissipates (the tail end),

---

attention to one of their various elements." (p. 99-100 of Niels Bohr, *Atomic Theory and the Description of Nature: Four Essays with an Introductory Survey*, Cambridge University Press, Reissue edition – 16 Jun 2011) with a focal point in between which corresponds to the current perceptual state of phenomenal experience.

This way of dividing the single conscious process in two parts with a boundary between them is somewhat artificial, but it can be very useful in understanding how it works and what is its significance for the evolution of matter in the universe. Viewed from that perspective, the conscious process is analogous to the familiar chemical reactions of sustained oxidation, like the burning of a candle, for example. This process also has two parts—a fuel source as input and exhaust gases as output, with the oxidation process continuously occurring at the boundary between them. In the same way the conscious state has an input in the form of sensory data and emerging internal conceptual states and output in the form of body movements and dissipating internal conceptual states. The conscious state and the qualia of phenomenal experience are situated at the boundary between these two parts, where the newly emerging entanglement resonances of particle structures in the brain clash with the already formed and gradually dissipating resonances.

It is a bit difficult to state in abstract terms what gets transformed in the case of the conscious process, unlike the chemical reactions, where one set of molecules gets transformed to another set, but we can note that the conscious process is of the same kind as the evolutionary process of life forms. In abstract state space terms it is at the heart of the evolutionary process, meaning that it also represents a process of accumulation of representations, but at a faster rate and with the highest density in comparison with the processes in non-conscious living matter. If we visualize it as a process involving particles in abstract state space, it would look like a dense particle structure in high-dimensional space which is very dense at the core and rapidly dissipating as one moves towards the periphery, or in other words, as a relatively tiny cosmic body made up of only a small number of particles. In the core of this body there are frequent processes of particle transformations, corresponding to the conceptual changes in our collective knowledge, while at the periphery the particle interactions are more regular and transformations are an exception (they would correspond to, e.g., a genetic mutation producing a new pathway in the network of biochemical reactions making up the living organism).

In more conventional terms, the continuous accumulation of representations at the heart of the conscious process can be understood as a process of continuous learning. Here we need to understand the word "learning" in broader terms, not just like what is going on when pupils attend school, but as the cumulative effect of the constant stream of perceptual and imaginational states which leave imprints in a person's memory and over time shape the knowledge and the personality of any human. Every sight and sound, as well as every thought and realization, leave a memory trace which physically represents a slight alteration in the complex structure of the shared entangled state, and over time these slight alterations accumulate and on average lead to a better representation of the world, i.e., the totality of all representations that make up the universe. This is what we refer to as the process of continuous learning or continuous accumulation of representations.

There is one more point worth mentioning regarding this process, namely, that the configuration of any instance of the unified conscious state as well as the superposition of all conscious states, as we imagined them in abstract state space, is unique at any given moment in time. In other words, the conscious states never repeat themselves. They constantly transform, but in a way that continuously produces novel configurations, and never go back to an exact copy of an old configuration, although most of the time they actually move to states very similar to older states (e.g., when we use a word which we know, which is nearly always the case) but never exactly the same as them. We will consider why this should be the case in the next chapter, but for now we should only note it as a proposition which will help us imagine correctly the workings of the mental state.

The Basic Thought Process (Reasoning and Concepts)

When we imagine the mental state as a physical entity realized inside the brain, we need to imagine particle structures spread throughout the brain and encompassing a substantial portion of the brain's mass, which are engaged in a shared entangled state. This state has some internal structure, which we can imagine as varying colors representing stronger and weaker entanglement, as well as connections between qualitatively different structures. The mental state transforms with time, changing its shape and colors, but it remains more or less in the same physical location, i.e., it always engulfs most of the brain. Its state, however, changes substantially, which corresponds to a movement along a trajectory in abstract state space.

This movement resembles jumps from one point in state space to another, since the mental state oscillates through successive phases of well-defined conscious states and unrealized transitions between them.

This behavior is typical for complex systems with chaotic dynamics, but unlike those made up of inanimate matter, the brain, as we said, always undergoes an evolutionary process of adaptation of its structures, never returning to an earlier state.

The constant switching from one well-defined conscious state to another is what we call reasoning, or the train of thought. Although it is chaotic, it is still a deterministic process, if we assume that all representations that make up matter behave according to a set of rules, namely, the known laws of physics applying to particles of matter plus eventually some undiscovered regularities. So, the trajectory of the thought process is dependent on the structure of the mental state at any given time and the impacts of the sensory data, i.e., influences from the immediate physical environment, on the current and future mental states. In that way it resembles the process of transformation of a particle state which depends on the internal representational configuration of the particle and the representational configuration of the particles in its environment with which it is engaged in an interaction.

Just like the particle transforms under the pressures from the representations in its environment, the mental state transforms under the pressures from its physical environment, relayed to it through the senses which represent this environment and in this way select and construct the images that constitute a partial and transformed replica of the environment. So, we need to imagine the mental state as having a component which is directly related to the environment, i.e., is a representation of it. This component is probably smaller than the one representing the experiencing, conscious self, and it is partly involved in the shared entangled state corresponding to the conscious self and partly realized by activity falling outside of the shared entangled state.

In this way the events in the environment (e.g., movements of objects causing changing patterns of light reflection and acoustic waves of oscillating air pressure) get transformed by the sense organs into altered states of entangled particle structures in the brain which in turn exert pressures on the main shared entangled state and guide its transformation. This representational relation is very strong and the link between the events in the environment and the pressures that transform the conscious state is pretty much automatic—when something happens in the environment we "see" it or "hear" it, i.e., our conscious state changes correspondingly without any conscious effort on our side. The conscious state appears to be guided in its transformation by the representational mechanism linking external events (i.e., rearrangements of particle structures, in the physical sense of the meaning of 'events') with the shared entangled state constituting the conscious mental state.

In a similar way the conscious state is guided also by pressures from the entangled particle structures inside the brain which fall outside of the main shared entangled state. These pressures are perceived as memories, unconscious instincts, rational arguments, etc. which intervene in the reasoning process and cause the conscious state to make certain jumps, i.e., to reach one particular state instead of another. As representations, these elements of the mental life are much closer to the main conscious state and it seems that the conscious state has some degree of influence over them, i.e., it can select and transform them to some degree in a mirror way to the pressures they exert on it. Still, they are not completely included in the main conscious state because they appear to consciousness as external factors modulating the reasoning process.

The reasoning process can be regarded as consisting of two types of acts—the accumulation of pressures from the environment of the conscious mental state through the mechanisms described above, and the sudden switch to a new configuration of the mental state, which phenomenally seems like a moment of realization or of learning of something new. Viewed this way, the reasoning process resembles the evolution of the particle states, which go through alternating phases of spreading-out of the wave function punctuated by sudden collapses to a physical state with well-defined measurable properties. Just like we can only observe particles in those well-defined states, we can only observe the mental state through concepts, i.e., words, that get communicated to us.

Indeed, if we try to imagine the configuration of all possible mental states situated in some kind of abstract state space, we will depict the act of reconfiguration of the mental state as a reconfiguration of the particle structure in abstract state space in response to pressures from the environment. So, making a realization and learning a new fact corresponds to an act of adaptation of particle structures in abstract state space. Thus, the main process occurring all the time in the brain and also in the microscopic realm of particles and particle structures coincides with the main evolutionary process of matter in the universe as a whole, which, as we saw already, is driven by the events of adaptations of particle structures to each other. The notion of reasoning turns out to be our way of conceptualizing this process of accumulation of representations, or at least the aspect of it relating to our realized mental life.

The acts of realization and learning new facts are also acts of conceptual change. Earlier we equated concepts with well-defined mental states realized by strongly coupled shared entangled states of particle structures in the brain. The act of a realization is an act of establishing a link between two mental states, i.e., two different configurations of particle structures which previously did not evoke each other, but after the conceptual change, due to the act of realization, become part of a single configuration realized by the global shared entangled state in the brain. In other words, the act of realization is a coming together of two temporally disparate concepts in a single moment in time, producing a new conceptual state with a more complex configuration. In abstract state space this event looks like the establishment of a representational connection between two (or more) particles, which joins particle structures in a larger shared entangled state (of higher dimensionality, since the particle structure in abstract state space form a very high-dimensional space). It is in fact an act of adaptation of the particle structures, much like the ones that are the hallmark of living systems in physical space.

The conceptual change described above would be classified as a generalization in the terminology of classical logic. In terms of particle structures in abstract state space this is an act of joining together of representationally entangled particle structures, producing a novel, more complex structure. It should be noted that the new structure is not a simple sum of the two old ones, but that the entire dynamics and degree of inclusion in the shared entangled state of the particles (corresponding to individual concepts) get transformed. This leads, for example, to a diminished role and degree of inclusion of the particles in the periphery, meaning that some of the properties associated with the old concepts wane in importance at the expense of some central properties or even some novel properties that may be created out of nothing. Similar phenomena occur with real physical particle structures, where the joining of two structures (e.g., molecules) may lead to reconfiguration of the bonds between the atoms, weakening some of them and strengthening others, and the creation of new resonances, which as we saw in the previous chapter are some sort of quasiparticles.

Besides the logical operation of generalization, also referred to as induction, there is an opposite operation called deduction, meaning that properties belonging to a general class of objects are ascribed to individual members of this class. In terms of particle structures in abstract state space this is again an act of establishing a representational link, but this time between one large structure (the more general concept) and a small structure (possibly an individual concept), i.e., it is analogous to the act of a large organic molecule acquiring an electron or a single atom, for example. In doing that, the shared entangled state resonance may dissolve in the main particle structure and shift to the newly acquired structure and from there to other structures, corresponding to shifting the focus of attention from the general concept to the specific property and from there on to other concepts.

In the same way, we can imagine all other logical operations that constitute our reasoning process as realized by transformations of particle structures in abstract state space corresponding to reconfigurations of the shared entangled state of real particle structures in the brain. These reconfigurations of the entanglement necessitate also small reconfigurations in the physical bonds among the atoms making up the brain mass. Since the brain is made up of adapted particle structures and processes, the small physical reconfigurations correspond to large shifts in the shared entangled state and they in turn can evoke through dedicated mechanisms large, macroscopic movements of other particle structures, namely the muscles and the body parts as a whole. This is how the conscious activity translates into physical actions of moving body parts, and vice versa, how macroscopic movements of objects in the environment perceived through the senses and the movements of body parts perceived through proprioception mechanisms impact on the conscious state. One particular aspect of this type of activity is the faculty of language, which involves a very close link between the conscious state and enactment and perception mechanisms, and it will be the next topic in our exploration of the brain and its mental life.

What is Language

Language is a nearly automatic ability. When we hear a certain type of sound, or see a certain arrangement of lines and curves, we can't help it but perceive this as a word. In this way, sounds and visual patterns evoke nearly automatically their corresponding conceptual states of entangled particle structures in the brain. The reason for this is that there are areas in the cortex where the neural structures are devoted especially to that purpose—to transform auditory and visual patterns into shared entangled states (also tactile sensations, in the case of blind persons) and vice versa, other areas 'read' the internal mental states and produce movements that cause sounds (in case of speech) or leave marks (in case of writing or typing).

Thus, language is a mechanism with which different physical instantiations of conscious mental states (i.e., different persons) can communicate and influence each other's state. Since the mental states are high-dimensional versions of particle states, both being representational complexes in nature, communication between them is analogous to communication between particles, i.e., what we call the force-carrying particles. So, exchange of a unit of linguistic information (i.e., a word) is analogous to exchange of a force particle, e.g., a photon. The analogy is rather deep—in both cases the state of the sender has to be similar to the state of the receiver for the exchange to occur. Also, the act of exchange alters the states of both parties, changing their natural course of evolution of the state and therefore their trajectories in state space. And finally, we can regard the units being exchanged as virtual, i.e., as epiphenomenal, and think of the process as a coordinated transformation of two entities linked by a representational relation.

That said, we need to note that the exchange of linguistic information is a much more complex process which never functions in exactly the same way as before, while the exchange of force particles is much more uniform and reproducible. We will examine the significance of this difference in Chapter 3 when we consider the totality of all representational states, which we call 'the universe'. For the time being we will consider in more detail the mechanisms through which the specialized neural structures of language processing modulate the conscious state.

We said that there are areas in the brain where the neural structures are specialized for language processing (in most people they reside in the left hemisphere of the brain). They are relatively small compared to the rest of the brain. This means that when we imagine the shared entangled state, most of its content would be devoted to representing a concept reflecting some state of affairs in the world (e.g., "chair", or "red", or "falls", etc.) by evoking the typical perceptual states that are associated with this state of affairs in the world, i.e., the visual percept of a typical chair, the visual percept of red or the visual percept of movement towards the ground, possibly associated with some sounds and subjective evaluative states, such as positive or negative feelings. In addition to that, a small part of the shared entangled state will be devoted to representing perceptual and motor states that co-occur with the perceptual state of the real object or action, such as the sound corresponding to hearing the word "chair" in a specific language, or the visual pattern of the written representation of the same word, or the movements involved in pronouncing or writing the word, etc. This part is the one realized by neural structures in the language areas of the brain.

It should be noted that the language part is present in the shared entangled state most of the time, except in activities without a linguistic component, such as continuous complex body movements as in, e.g., dancing or sports, or continuous perception of such activities, e.g., watching sports or listening to music.

We also said that the sustained existence of the shared entangled state is due to the high density of adapted particle structures and processes (in terms of cascades of chemical reactions) in the brain. The constant activity and rearrangements of these adapted structures, which is essentially the normal activity for sustaining the life of biological organisms, is what makes the shared entangled state transform constantly in a sort of pulsating fashion—jumping from one well-defined state to another with brief transitions of lesser degree of coherence and entanglement among the structures. In this way, the well-defined stages of the shared entangled state have the potential to represent both static and dynamic phenomena in the external world. The static ones are represented through a well-defined state corresponding to a steady perceptual activity, such as the seeing of a uniform, unchanging red color. In this case the whole system, consisting of the object in the external world, the sensory organs and their corresponding areas in the cortex, and the rest of the cortex, realizing the main part of the conscious mental state, are briefly engaged in a steady state (on the order of a few hundreds of milliseconds), corresponding to the representation of the static phenomenon. The dynamic phenomena are represented by a well-defined resonance with approximately the same duration, but in this case the main part of the conscious state is transforming as well, i.e., it does not represent a static percept, as the seeing of red color, but a dynamic percept, as watching the fall of an object towards the ground.

These two types of representational states of the conscious mental state are mirrored in language by the two major categories of words—nouns and verbs. These are open categories, meaning that new members are constantly created and lost, reflecting the fact that the abstract conceptual structures of our collective knowledge are constantly changing, with new concepts being created and some old ones being forgotten.

In addition to those two categories of words, referred to as parts of speech in the linguistic terminology, there are also other major parts of speech with somewhat smaller numbers of constituents and secondary role in grammatical structures, namely, adjectives and adverbs. The adjectives are usually modifiers of the meaning of nouns, while the adverbs typically modify the meaning of verbs but can also apply to some phrases and adjectives. For example, "red" is an adjective and it can apply to the noun "chair", while "smoothly" is an adverb and typically applies to verbs, such as "falling", but can also apply to adjectives ("smoothly red color").

In terms of the shared entangled state, a modifier is a well-defined state which could represent the main conscious state, but could also be a secondary state, activated partially together with the main state. So, it is in a way an additional structure attached to the main structure of the shared entangled state, i.e., it consists of particle structures engaged in a shared entangled state, which are probably fewer in number than those realizing the main conscious state, and are also different physically from them, and which join the main shared entangled conscious state, thus adding to its structure. Such secondary states can supplant both a static and a dynamic well-defined state, so they are also static or dynamic in nature, constituting adjectives and adverbs, respectively.

Besides those main categories of noun—and verb-like parts of speech, there are also many other parts which guide and modify the evolutionary process of the mental state in some way. For example, there are particles, like "to" or "as", the definite article "the", pronouns like "this", auxiliary verbs ("have" and "is" in English), etc. All those parts of speech can be imagined as performing some modifying function on the mental state. For example, the particle "to" can be said to spur the evolution of the mental state towards the next well-defined state, analogous to a push applied to an object moving towards a target, the definite article "the" seems to boost the retrieval from memory of the most familiar item currently activated, etc.

In order to develop a full inventory of the action of each part of speech one would need to have a complete picture of the whole phenomenology of the functioning of the mental state in the brain, which is beyond the scope of the current book, although many of the main phenomena are elaborated in some detail here and could help in that respect.

In summary, we can say that there is something like a 'language module' in the brain—specialized areas of the cortex whose shared entangled state of the particle structures is an integral component of the global shared entangled state mediating the conscious mental function, which serves the purpose of converting the physical representations of external sensory stimuli (e.g., spoken or written words) into modulating influences exerted upon the evolution of the conscious state, and vice versa, reading the conscious state and producing linguistic stimuli (speech or writing). The individual linguistic units (words or phrases) correspond to well-defined conscious states and/or the propensities of their transitions to other states (as in "to"). Thus, language serves the purpose of mediating interactions between individual conscious states, much like particles of forces mediate interactions between particles.

Since the language component is an integral part of the conscious state, which makes language processing an almost automatic, unrealized activity, we can regard the collection of individual thought processes of all people in the world as a loosely connected network of a collective thought process, although physically the conscious states are not connected in a shared entangled state. This situation is analogous to the way the individual locally entangled particle structures in the brain form one unified shared state—they do this via non-entangled (we can also say 'classical') channels for communicating influences. Those channels are much more dense and with higher bandwidth among the particle structure in the brain, and much more limited in the case of linguistic communication among people (i.e., the conscious states realized in their brains), but nonetheless, there is an analogy between the two cases.

From the broader perspective of the fundamental process of accumulation of representations in the course of the evolution of matter in the universe, we can say that language makes consciousness more widely spread—not confined just to individual brains, but spread out in a loosely connected network of brains. This results in better accumulation and transmission of knowledge (i.e., representations) from one person to another, which is ultimately in line with the fundamental process of accumulation of representations of matter in the universe.

What is Memory

We postulated that the conscious thought process represents a constant switching from one well-defined state to another. These well-defined states have configurations which are highly similar to earlier instantiations of themselves, but never exact copies of the earlier instantiations—always containing an element of novelty. The high degree of similarity allows us to talk about them as instances of the same thing, i.e., of the same (in some abstract sense) shared entangled state, which we call a concept. This "sameness" to some degree of approximation is also the basis of the phenomenon of memory.

When the mental state switches from one well-defined state to another, occasionally it switches from a less rehearsed state to a more rehearsed state (i.e., one with higher incidence of earlier instantiations). In that case, the mental state switches from a state of looser entanglement to a state of higher degree of entanglement and coherence. Bringing up our picture of the two components of the mental state—the front end, which is constantly being constituted, and the tail end, which constantly dissolves—we can think of this case as a movement from a poorly constituted, easily dissolving tail end to a well-constituted, easily aggregating front end. If this is the case, then for a moment the entangled state would expand into the new well-prepared entangled structures of the front end, and this would feel like a perception event (in perception events the states of the cortical areas connected to the sensory organs form the well-prepared entangled structures into which the mental state expands). In the absence of a strong sensory stimulus driving the formation of the well-prepared structure, the experience will be vague compared to an actual perceptual experience, and this kind of experience is what we conceptualize as memory recall.

For example, a smell or a color experience (seeing red) may push the mental state into a well-defined state where the smell or the color experience played a central role in an earlier perceptual event, bringing all other associated structure into the shared entangled state and effectively invoking a different conceptual state (that of seeing a rose). In that case, there will be no perceptual event of seeing a rose based on the sensory data, but there will be features of the mental state corresponding to seeing a rose briefly engulfed in the shared entangled state (since it is well rehearsed, or well prepared because of the idiosyncrasies of the neurochemistry processes in the brain at that time), and this will be noted as the phenomenal experience of a bursting in memory of a rose.

In this way the nature of memories is closely related to the nature of concepts—both can be basically understood as instantiations of well-defined states of the same kind at different moments in time. Memories can also encompass entire episodes containing multiple concepts or even sequences of events, but the basic principle is the same—re-instantiation of earlier well-defined states.

Since the act of remembering represents a return to an earlier state, it also serves to reinforce the conceptual structures that already exist. In terms of the picture of concepts forming a high-dimensional structure in abstract state space, the returning to earlier states has two-fold implications. First, it gives the conceptual structure its very existence, i.e., we can imagine a concept as a particle in abstract state space precisely because of the similarity of the well-defined states representing the different instantiations of the concept. Since the physical configurations of particle structures involved in the different instantiations are very similar, the location of the concept in abstract state space is more or less the same, which gives it some permanence, and therefore, existence.

The second implication is the fact that the concepts tend to be conservative with respect to change, i.e., they like to stick to the old location in state space. As we saw in chapter 1, this is analogous to the inertia of particles due to their mass. In chapter 1 we considered the force of gravity as a kind of memory of the representational content of particles for their earlier states and interactions. The analogy goes the other way too—we can imagine memory as a kind of epiphenomenal force in abstract state space which keeps the particles in place and close to each other in compact conceptual structures. This force arises from the fact that conceptual states are all dependent on each other and do not exist in isolation. Thus, any transformation in one concept results in coordinated transformations in other concepts, keeping them bound close to each other in state space.

Most of the brain activity is reinstantiations of already existing concepts but in novel contexts, constantly producing novelty in the conceptual structure as a whole. Occasionally this novelty results in rearrangements of the particles constituting the conceptual structure and even the appearance of new concepts. Thus, throughout the most part of the course of life, the conceptual structure accessible to a single individual's mental state grows continuously, expanding also the range of memories that can be recalled by that individual. As we said earlier, the mental state always has a current state, which is one particular location in the conceptual structure, and it has a propensity to move, or jump, to a new (usually neighboring) location. When that happens in a well-prepared manner, the experience is one of a recollection of a memory. In this way, a larger and more diverse conceptual structure translates into a larger potential for experiences of memory recall to occur.

The mental state, pictured as occupying a particular location in the conceptual structure in abstract state space, engenders roughly three distinct regions in the conceptual structure: 1) its immediate location, which includes one central, currently active concept and a few related ones which are partially active, 2) its immediate neighborhood, which includes all possible conceptual states to which it could jump, and which influence to some degree its evolution, and 3) the rest of the conceptual structure, which is inactive at the given moment. These three arbitrary regions correspond to the three types of memory identified by psychologists: the so-called sensory (or iconic), short-term (or working), and long-term memory.

The sensory memory is considered to be the most vivid and detailed type of memory, but also the shortest-lasting one, with a duration of under a second. This is the time interval during which the immediate perceptual state can be retained and exploited in an experimental task. In our terminology this would correspond to using a part of the currently active mental state, i.e., a part of the currently entangled particle structures, constituting the current perceptual state, to affect another part of the currently active mental state, possibly the part which is just emerging and being included in the shared entangled state, and which constitutes the experimental task (usually conveyed through language, e.g., "recall the red shape and tell me how many vertices it has!").

The short-term, or working, memory has an intermediate span of 10-15 seconds, and it also has an intermediate retention capacity. The typical example is holding a phone number in memory while dialing it. It can retain fewer features from the perceptual or the internal conceptual state that needs to be retained, but considerably more than what can be retained for a longer period of time in the long-term memory. From the point of view of our terminology of conceptual structures in abstract state space, these are structures that are close to being currently active, i.e., they do not correspond exactly to the current entangled particle complex, but are close to it in their configurations, and therefore it is possible for the current entangled state to switch to any of these potential configurations. This allows them to be retained partially active and to transition to them in the scope of 10-15 seconds, like in the case of retaining a 7-digit phone number.

The long-term memory is whatever is retained beyond the extended time frame of the currently active mental state and it is believed that physiologically these memories are mediated through the mechanisms of long-term potentiation (LTP) and long-term depression (LTD) of synaptic transmission. These mechanisms alter the coupling strength of synapses, changing the probability that the postsynaptic neuron will fire when the presynaptic neuron fires and triggers the release of neurotransmitters in the synapse. There is evidence that this mechanism operates in an anatomical substructure in the brain called the hippocampus, which has been implicated as a crucial component in long-term memory formation. Damage to the hippocampus bilaterally results in inability to remember the past beyond the time frame of the current mental state (i.e., the last few minutes) from the moment of the occurrence of the damage. Patients with this kind of disability live in a constant present, as if they have just recently regained consciousness and remember events from the past only from the time before the damage. The memory consolidation orchestrated by the hippocampus is believed to be particularly efficient during sleep. Anatomically, the hippocampus seems to contain a map of the different cortical regions with connections between them and the corresponding areas in the hippocampus.

It is probably not surprising that long-term memories are realized by significant rearrangements of particle structures, which is what the mechanism of long-term potentiation actually represents. If new mental states are to be formed, that would necessary require some rearrangements of particle structures, since by our hypothesis this corresponds to changes in the configuration of the shared entangled state. The fact that it happens at the synapses is also in line with our hypothesis of what constitutes the physical manifestation of the conscious state—the particle structures that constitute the shared entangled state are typically located inside the neurons, and the synapses, which determine the properties of the communication among the particle structures, are the suitable locus for modulating their interactions and therefore the configurations of the shared entangled states instantiated in a given time stretch. Those configurations precisely constitute the thought process and the memories being recalled during the thought process.

In this way, we come to regard memory as a physical process which is actually one aspect of the basic thought process, namely, the fact that the basic thought process nearly always re-instantiates earlier mental states with some minimal added amount of novelty making each instantiation unique. For that reason, memory does not have a specific location in the brain, but encompasses the whole brain, just like the conscious mental state can potentially be realized in any brain structure. The experience of remembering an earlier mental state depends on the degree of difference between the coupling strengths of the current shared entangled state and the future one, i.e., on how well the future state is prepared and how readily the transition to it occurs. The preparation of the future state occurs spontaneously in particle structures which are not part of the current shared entangled state and they will be the subject of the discussion in the next section.

What is the Unconscious

We said that the conscious state is realized by the shared entangled state of particle structures which are functionally similar at some given moment in time, i.e., are situated roughly in the same location in the abstract state space of the possible regimes of physical oscillatory states that they could possess. This shared entangled state is a dynamic entity, constantly engulfing new structures which are prepared in some way, i.e., their states converge towards greater coherence, and also constantly releasing structures whose coherence dissipates. Thus, the shared entangled state probably occupies only a small proportion of all matter making up the brain, not more than half of it at any rate and more likely much less than that. The rest of the activity in the brain is for the most part just other metabolic activity needed for sustaining the life processes in this organ of the body and part of it is also weaker entangled resonances which are different from the main one of the conscious state and therefore do not belong to it. Those resonances are namely what is classically considered as the unconscious (or subconscious) part of the thought process.

The unconscious entangled resonances have the possibility to influence the evolution of the conscious state directly if they get included in the subsequent well-defined state, or indirectly if they do not get included, but they affect the degree of coherence of another conceptual component which is included in the conscious state and in that way the effect it has on the evolution of the state.

In terms of our picture of the conceptual structure in higher-dimensional abstract state space formed by the different possible mental states, the unconscious resonances would look like partial activations of neighboring concepts and the nearby space around the currently active elements corresponding to the conscious state. This picture is in fact very much like the picture of virtual particles we encountered in chapter 1. Just like the virtual particles, the unconscious resonances are not directly observable by the conscious state but they influence it in some ways. Also, they represent different possibilities that the evolutionary trajectory of the mental state might take, just like the virtual particles represent all possible ways the particle state may evolve from one particular state to another. The trajectory eventually picks out one of those possible paths, thus deciding on what the future state will be, collapsing the multiple possibilities into one realized actuality.

So, we need to imagine the unconscious brain activity as some kind of haze surrounding the conscious mental state, as we imagine it in abstract state space. One corollary of that picture is the observation that the unconscious extends beyond the boundaries of the conscious state, i.e., has larger extension and therefore encompasses conceptual elements that would otherwise be left out of the shared entangled resonance of the conscious state and would be unrelated to each other. Occasionally, the newly formed, well-defined conscious state includes two or more such elements, which previously were in the unconscious periphery of the earlier conscious state, and this event is perceived as an act of a realization, or establishing a connection between distant (but related in some way) facts, which is in fact what we conceptualize in retrospect as the working of the unconscious part of our minds.

For example, we may wonder about the solution of a problem, e.g., trying to understand the flow of electricity in wires as described by Ohm's law, and we may realize after a while that it resembles the flow of water in pipes. That realization is an example of analogy making involving the act of bringing unconscious elements into consciousness. The analogy is between the flow of electricity and the flow of water, and initially the concepts of electricity and water are unrelated, i.e., when we are in the state of wondering about the problem they are both outside the conscious state, but when we gradually build the conceptual framework of the imagination of how electricity flows, at some point this framework merges with the one representing our existing knowledge of the flow of water in pipes, and at that moment both concepts are brought into the conscious state and a connection between them is established.

The ability to make realizations, including those based on analogies, is generally taken as a sign of intelligence. Indeed, such events result in the creation of new knowledge, and since the majority of our knowledge is derived from experience, it represents the external world to some degree of approximation, so insights usually result in creation of truthful knowledge about the world. In terms of the conscious state and its accompanying unconscious surroundings we can say that the events of realization depend on the breadth and the balance among the relations in the conceptual framework included in the conscious and the unconscious parts. In other words, the larger the extent of the conscious state and its corresponding unconscious surroundings, the more concepts are being considered at any given moment and the more likely it is for the cognitive process to take the right route and to reach the correct decision. Also, more balanced weights of the links among different concepts in the framework translate again into more alternatives being considered and higher likelihood for taking the optimal decision. On the contrary, more entrenched relations mean that the evolution of the conscious state is to a great degree predetermined, meaning that in cases when the optimal decision is not the obvious one, it will be missed.

These two properties of the conceptual framework—the breadth of the mental state and the weight of the connections among the concepts—are in fact related. If we assume that the mental state has some limited volume in terms of conceptual space in abstract state space, then in the case of more balanced connections it would be able to spread further along those connections from its focal point, meaning that its extent would be larger. On the contrary, less balanced connections would result in most of the volume extending along one or a few of the connections, making the state more concentrated.

We can imagine this proposition also in terms of the resonant entangled state of particle structures in the brain. If they tend to resonate in one specific way, then the tiny influences from the unconscious part of the brain activity would have little effect. If, on the other hand, the shared entangled state consists of diverse resonances of similar coupling strength, then the influences from the external unconscious resonances would be significant, effectively including an even larger number of conceptual states in the main conscious state, making the decision making process more informed and consequently more successful.

The degree of balancing of the conceptual connections is a matter of habit which can be imitated by one person from another and in that way becomes culturally transmitted. Different social and ethnic groups may develop different habits regarding the basic cognitive skills of paying attention and decision making, resulting in different levels of acquired intelligence and consequently social status. These differences may ultimately be traced down to the workings of the unconscious and its influence on the properties of the conscious state. A richer and more balanced unconscious would result in more analytic thinking and more careful decision making, enhancing the ability of people to manipulate the world around them, which in economic terms means greater wealth.

What is Sleep

The unconscious activity in the brain is unrealized, but it is still part of the normal functioning of the brain in the awake state. There is also another regime of operation of the brain which lacks consciousness, namely, the sleep state. It occupies about one third of the daily cycle, and during that time the brain is functionally in a different mode of operation altogether. Very little is known about the neurochemical processes at work during sleep and very generally about the biological purpose of sleep. We can only say that it is very important for the normal functioning of the body, since the deficit of sleep, which occurs in some rare brain disorders, eventually leads to death after a few weeks of sleep deprivation.

The leading hypothesis in modern science about the purpose of sleep is that it performs some kind of regeneration or replenishment of neurochemical resources, i.e., it restores the normal balance of substances in the brain, which presumably gets gradually disbalanced in the course of the awake activity during the day. A similar phenomenon is observed in artificial neural networks—in a learning task initially they learn to perform the task better and better, but eventually they start to 'overlearn', the connection strengths between the nodes settle in a stable configuration and become more and more entrenched, leading to worsening performance on novel tasks. This phenomenon is considered to be an abstract model for what is going on in the brain—during day time the synaptic connections are supposedly receiving more and more training, making them more committed to their specific function, and this tendency needs to be counteracted by some sort of 'recalibration' or 'normalization', which is what happens at night during sleep.

The regeneration hypothesis about the function of sleep also makes sense in the framework of conceptual structures in abstract state space representing human mental activity. We saw that we can think of individual mental states as particles in a high-dimensional abstract state space which aggregate under the influence of an epiphenomenal force analogous to gravity. This force tends to compress the conceptual structures with the consequence of fixing them in rigid configurations. This phenomenon is equivalent to the overlearning in artificial neural networks, and it needs to be counteracted by some relaxation of the coupling among the conceptual mental states which will allow them to move and reconfigure, i.e., it would allow for new learning to occur. During sleep it is possible for a different set of metabolic processes to operate and to reduce the coupling strength among particle structures by restoring the usual balance of biochemical substances in the brain.

It is clear that the regenerative mode of operation of the brain is very different from the learning mode of the awake activity. During learning particle structures need to form entangled ensembles leading to consciousness experience, as we discussed earlier. Regeneration, however, is a local operation, acting differentially on individual particle structures, so the entanglement, and consequently the unified conscious state, is lost. It is also possible for the brain activity to be in intermediate states between full-fledged consciousness and total absence of consciousness—these are the states of dreaming.

Dreams, in this way, turn out to be simply an intermediate ground between the awake and the sleep modes of operation, combining features of both of them. They still involve entanglement among particle structures, but it is a weaker one compared to the awake state, since the dream state is closer to absence of consciousness. The weaker entanglement allows for more structures to get included in the shared state, making it more extended and encompassing regions in abstract state space that normally would fall outside of the conscious state and would belong to the unconscious periphery or beyond.

This observation offers a good explanation for the unusual logic of dreams, noticed long time back by the founders of the mental health practice of psychoanalysis. Dreams seem to combine elements which are usually unrelated in the awake conscious state, but close to being related, i.e., residing in what we described as the unconscious (or subconscious, in psychoanalytic terms) periphery of the conscious state. Thus, they fall outside of the more concentrated conscious state (based on stronger entanglement coupling of particle structures) during the awake state, meaning that we normally do not notice the connection between these more distant concepts. The more spread-out nature of the dream state includes them in a single mental state, and if we happen to remember the dream after we get awake, we notice the connection between them, which is a novel experience for us and we find it unusual.

In psychoanalysis this property of dreams to relate more distant concepts is referred to as the symbolism of dreams. Symbols are in fact individual concepts isolated from a given conceptual scheme, representing in the awake state one or several closely related conscious states. The symbolic nature of dreams results from the phenomenon of bringing together in one conscious dream state symbols from different conceptual schemes. This cannot happen in the awake state because those symbols cannot be activated at the same time due to the fact that they belong to different conceptual schemes (by the very definition of a conceptual scheme, which, as we mentioned above, represents a set of conscious states that can be co-active). In dreams, however, the more spread-out nature of the conscious dream state can, and indeed has to, include concepts belonging to different schemes, leading to the peculiar logic of dreams that psychoanalysts call symbolism.

Another notable difference between the awake and the dream states concerns the nature of the self, or of one's self-awareness. While the self is strong and dominant in the awake state, it seems to dissolve and nearly disappear in the dream state, reducing one's conscious perception in that state to the level of childhood experiences or even a weaker self which is just observing and reacting to events in the dream but not able to act or to think consequentially. In other words, consciousness in the dream state lacks free will or goal-directed behavior.

These properties of the conscious dream state can also be explained within our framework of the shared entangled state of particle structures. Earlier we observed that the shared entangled state can be arbitrarily divided in two parts: the front-end of the newly constituting coherent ensembles of particle structures and the tail-end of dissolving coherence of particle structures. The boundary between them is what we postulated to be the locus of conscious experience with its associated perceptual qualia. This boundary is also the locus of what we regard as the self. Intuitively, we think of the self as the active conscious awareness of the world and our inner mental life at the very moment of the present state, exactly the moment we call 'now'. But the awareness at this moment can be physically located only at the boundary of the constituting and dissolving parts of the shared entangled state within our ontology of consciousness. Therefore, the physical manifestation of the self is exactly this boundary, and the properties of the self as e.g., self-observing or self-aware and aware of the world, must be sought in the particular properties of the state of matter at this boundary, namely, entanglement among structures of physical particles, i.e., representational complexes constituting with their behavior three-dimensional space and the familiar laws of physics.

If we accept the proposition that the physical manifestation of the self is this boundary, then it becomes easy to understand why the self is weakened in the dream state. Given that the dream state represents a state of weaker entanglement among particle structures, this means that the boundary between the two arbitrary parts of the conscious dream state will be smeared, i.e., there will be no clear distinction between the front-end and the tail-end of the conscious state, but the two parts will be gradually transforming into each other. This corroborates also our hypothesis that regeneration is a local operation acting differentially on individual particle structures thus making the entanglement among them much weaker or absent altogether. A gradual transformation of the coherence among particle structures allows for local variations and therefore for regenerative operations to take place, while a sharp boundary requires strong coherence which would prevent any local variations.

Finally, we need to note that sleep is in fact an altered state of consciousness, and as such it has some relation to other altered states of consciousness, such as those induced by psychedelic drugs or naturally occurring in the form of mental illnesses. Normal dreams resemble the hallucinations experienced under the influence of some drugs or in conditions such as schizophrenia, suggesting that in those states the brain enters an abnormal regime of functioning and is forming shared entangled states which are grounded in an altered balance of the typical biochemical reactions needed for sustaining life. Since the sleep state and the awake state are the two normal regimes of interrelated biochemical reactions, if the brain strays away from any of them it would affect also the mental life of the individual, which is what we regard as a psychic illness. The only two logical possibilities for the mental state is to become either more concentrated (stronger entanglement) or less concentrated (weaker entanglement), and in the latter case the abnormality would resemble the dream state of consciousness.

What are the Senses

The senses represent gateways through which external objects and events interact with the conscious mental state via some sequence of representational transformations of the signals originating in the environment. The main senses are vision, hearing, smell and taste, touch and temperature sensation, and there are also some other distinct sensory systems, like those for balance and orientation regarding the direction of the force of gravity, for pain sensation, and in animals for detecting electric and magnetic fields, pressure variations, etc. What is common to all these systems is that they relay information about the collective behavior of particle structures external to the organism, such as emission of photons in the case of vision, generation of air pressure waves through vibration in the case of sound, emission of chemical substances in the case of smell and taste, changes in temperature and pressure, mechanical interactions in the cases of touch and pain, and so on.

The events in the environment are typically very frequent—light is constantly reflected by objects during daytime, sounds are constantly occurring, chemical substances travel through the air continuously, and so on. They are also correlated with the behavior of the objects and especially aspects of it that are significant for the organisms, such as signifying the presence of food, mates or danger in the vicinity, or simply carrying information regarding what is where and how it moves. These correlations are very useful for the cognitive apparatus of the organism as it tries to build a representation of the world around it, and they shape the structure of the shared entangled mental state and its evolution. For example, the pattern of light impinging on the retina at any given moment produces a pattern of concurrent neural activity in the retinal cells, which gets translated and transformed to a more complex pattern of topographic representation of this neural activity in the activity of the neurons in primary visual cortex, which in turn gets transformed to an even more abstract and more complex neural activity representation in secondary visual areas, and eventually it affects the patterns of activity in the rest of the brain and therefore the shared entangled state. The story is similar, although the processing is simpler and with fewer stages, in the case of the other senses.

We can think of this cascade of representational transformations as transformations of the signal originating from the external objects and events. In this context, we need to understand the notion of 'signal' as an arrangement of particle structures and their movements which carries information about the particle structures at the beginning of the causal chain of transformations. In this sense, the idea of 'signal' becomes analogous to the one of 'representation', which is already familiar to us.

So, the signal generated by the external particle structures and their movements gets transformed and relayed all the way to the particle structures participating in the shared entangled state, and the intermediate particle structures through which the signal passes is what we call the sensory systems, or simply the senses. These are usually dedicated structures, like the ones making up the sense organs and the primary sensory areas in the cortex, which are not involved, or only very weakly involved, in other activities. In that way, we can think of them as cascades of representations at the interface between the external world and the mental state, in contrast to all other particle structures making up the organism, which are mostly inward-directed representations of other structures in the organism (by virtue of being adapted to them, e.g., the different anatomical organs being adapted to the rest of the organism by performing a specialized function necessary for sustaining life).

Thinking of the organism as a concentration of representations, physically realized as a giant assembly of particle structures in constant flux, allows us to conceptualize not only how senses are different from the other functions of the organism, but also how they came into being in the course of the evolutionary development of the organism. The difference between the particle structures participating in the sensory systems and those in the other organs of the body is that the former serve the purpose of transforming representations in a specific way along multiple stages, while the latter perform simpler transformations, usually in a single stage (i.e., a single metabolic pathway).

The development of multiple stages of representational transformation adapted to each other is a more complex process than the development of a single stage, but it still can be adequately explained in evolutionary terms. Take for example the evolution of the eye. It occurred over hundreds of thousands of years and it happened independently in different organisms many dozens of times. The evolution from light-sensitive patches to the complex vertebrate eye went through multiple stages of added complexity and structure to the organ, starting from a light-sensitive layer of cells (a proto-retina) projecting axons to the brain, folding later into a cavity with a narrow opening which had the effect of producing an image of the visual field on the retina, filling the cavity with a transparent gel-like substance, developing a shielding at the opening, which served also as a lens which focused the light rays, and eventually developing an iris and muscles which can focus the lens at different viewing distances. This sounds like a very complex process, but nevertheless, it is a logical sequence of accumulation of different adaptations with highly significant survival advantage for the organism, which led to a fast evolutionary process driven by natural selection.

Thus, the evolution of the complex mechanisms for transformation of the representations of external events reaching the surface of the body can be explained in evolutionary terms as accumulations of adaptations (or representations, in a more abstract sense), with a highly significant evolutionary advantage in terms of the survival of the organism, but also as a process of concentration of the accumulated representations, according to our picture of abstract state space, connected with the highly concentrated core of representational structures constituting the conscious mental state realized by entangled particle structures in the brain. From that perspective, the senses are extensions of the representational core of the conscious mental state, reaching out and linking it with the less concentrated representations in the environment.

Vision

The cascades of representational transformations in the visual system have been studied extensively and some of their more basic functionality is already known to modern science. We know that photons striking certain proteins in the retinal cells may evoke neural firing which reaches primary visual cortex and in turn evokes local patterns of neural firing activity there. Already in the retina there is some integration of the visual information, since the retinal cells do not simply respond linearly to the intensity of the incoming light, but they respond also to the local properties of the pattern of light, e.g., whether it is bright in the middle and dark in the periphery, or vice versa. In the primary visual areas the neurons fire preferentially to lines and edges at different orientations, while in higher visual areas they respond to more complex features, like corners or curves of different shapes. Ultimately, there are areas in the brain where a neuron might fire in response to an abstract concept, such as the identity of one particular person or object.

We need to make a brief digression here to discuss the meaning of the increased firing rate of a neuron, or a group of neurons, which can be detected with modern imaging methods like electrophysiological recordings and fMRI. The generally accepted interpretation in modern science of the increased firing rate of neurons is that in this way they process more information than in the rest state, analogous to a microprocessor which lets or stops current flow at irregular intervals compared to its rest state of a strictly cyclical activity with no informational significance. Thus, the informational significance of the neural firing events lies in their frequency and relative temporal asynchrony.

According to our hypothesis of the nature of the conscious state, however, the neural firing events are just a mechanism for modulation of the shared entangled state, which is the actual physical embodiment of the conscious mental state and not the neural firing events. In this mechanism the firing events have the effect of slightly altering the state of the neurons receiving the excitatory or inhibitory signals caused by the action potentials, thus making it more coherent or less coherent with the shared entangled state. An increased firing rate would simply make this mechanism more effective, i.e., the neurons would be more responsive to the influences from other neurons, meaning that their states would be more tightly coupled.

If we think of the shared entangled state, representing the mental state, as possessing some structure, i.e., if we think of the content of a given thought at one particular moment, we can picture it as being represented in the shared entangled state by resonant particle structures which also have some differences, but remain entangled because there is still a high enough degree of commonality in their states. In other words, the particle structures making up the shared entangled state would not be in exactly the same state, but they will have some differences which would be gradually transforming with some sharp boundaries delineating the structural components of the entangled state. (This picture refers to the mental state in abstract state space, not to the actual physical distribution of particle structures in physical space.)

We also pictured the mental state as being a representation of the external particle structures in the environment or of the internal particle structure coherences formed in the brain which fall outside the main entangled state. In the case when this representation goes out of synch with the transformation of the particle structures it represents, it will be hard-pressed to synchronize again with them in order to remain a faithful representation. In reality, this cycle of synchronizing and desynchronizing of the representation with its environment is constantly occurring, and this is basically the constant pulsation of the conscious mental state (and of the particle state as described by quantum physics), going through alternating stages of a well-defined, concentrated state and a fuzzy and smeared state.

Since the mental state represents its environment, its evolution parallels the evolution of the state of the environment, so it also anticipates the changes in the environment to some degree. When the changes in the environment diverge more from the anticipated course of events, there is more mismatch in the different parts of the mental state representing the different components of the environment, meaning that the different parts of the shared entangled state would struggle more to achieve a reconciled state with each other. In physical terms, this would require stronger coupling between the particle structures representing the different parts of the shared entangled state, which by our hypothesis regarding the rate of neural firing would entail a higher rate of firing, and this can be detected by the imaging methods.

Thus, the higher rate of firing occurs in neurons which are at the boundary of different representational sub-components of the mental state and which get out of synch because the environment that they represent changes in unexpected ways. This constant desynchronization and synchronization of the coupling of the components of the shared entangled state forms the boundary between the newly constituting and already dissolving parts of the mental state that we envisioned earlier, and this boundary is what we regard as the locus of conscious perception, or the awareness of the present moment. So, the brain areas with heightened neural activity that 'light up' in fMRI recordings ought to be more involved in the instantiation of the mental state, which is in line with the conventional interpretation of fMRI findings.

With regard to visual information processing in the brain, the neural activity in the primary visual areas in the cortex increases due to either changing visual patterns (temporal variations in the signal) or spatial light intensity transitions, i.e., edges, (or spatial variations in the signal). Thus, mapping of the topographic representation of the visual field in primary visual cortex with fMRI recordings is usually done with flickering checkerboard patterns in order to maximize the neural activity in the cortical regions that are homologous to the location of the pattern in the visual field.

This property of the response of the neural structures in visual cortex to external stimulation is unique for this area of the brain, leading to a specific organization of the connections among the neurons and their topographic distribution. In primary visual areas the neurons cluster in small patches with similar responses of the neurons to external stimulation from a particular location in the visual field and with sharp boundaries between the patches. This organization is still close to a veridical topographic mapping, but it also goes some way towards integration of information and formation of higher-order representations (e.g., of basic visual features and shapes).

In general, we can picture the visual system as a hierarchy of abstract layers, forming a pyramidal structure, where each successive layer integrates information from the preceding one, thus transforming the two-dimensional signal of the image of the visual field to an abstract representation of a region of space populated with objects, which is what ultimately partakes in the conscious state. The processing of a novel stimulus moves like a wave from the bottom of the hierarchy to the top in the timeframe of a few hundreds of milliseconds, but the ultimate percept available to consciousness involves coherent activity from all layers, i.e., all cortical visual areas. If we think of those layers as neural networks, we can say that the lower layers settle somewhat earlier than the higher layers, but in the end state they all settle in one unified representation of the objects present in the visual field.

The unified representation persists briefly, for less than a hundred milliseconds usually, and then the process repeats again. This is the familiar pulsation of the mental state going through well-defined and fuzzy stages. It is paralleled by the pattern of eye movements—the eyes constantly jump, or saccade, from one location in the visual field to another, with brief fixations in between. The fixations are the time periods of a stable image formed on the retina, allowing the visual system to settle in one coherent representation, and during the saccades the visual processing is suppressed by a special neural mechanism preparing the visual system for the next percept. Those successive fixations constantly update the representation of the environment in the higher-order areas in the brain, allowing us to stay aware of the space and objects around us. Most of the time the environment is relatively unchanging, with very few objects moving in simple ways, so our representation of the environment is close to complete. In some special circumstances of a rapidly changing environment, such as while playing a team sport, we can only track some of the changes, e.g., the movement of the ball and maybe one or two other players, but our brains cannot build a complete representation of the environment.

The unified representation settled in a well-defined state is something like a miniature model of the physical environment that contains some of the properties of the objects situated in it, like the shapes of their surfaces and the ways they reflect light, their likely movements, etc., but not others, like the inner structure of the objects or the chemical transformations taking place inside them. This model encompasses a representation of three-dimensional space and movements and interactions among macroscopic bodies in such a space, which reflects the properties of physical space outside the brain and the behavior of the macroscopic objects situated in it. This means that the shared entangled state of particle structures has the adequate internal structure to form such representations of three-dimensional space.

On the other hand, earlier we saw that the three-dimensionality of physical space arises as an epiphenomenon from the way the representations making up the physical particles interact with each other, which in turn depends again on the inner structure of those representations. Thus, the inner structure of matter, when viewed as representations, is responsible both for the behavior of the particles which forms a three-dimensional space, as well as the three-dimensional representation of space and the corresponding behavior of macroscopic objects, formed inside the human mind. This is probably not just a coincidence and it suggests that the awareness of space and the objects inside it is formed at the level of the inner structure of the particles engaged in the indirect entanglement of the shared entangled state. This representation residing in the microscopic inner structure of particles, however, requires an organization of the particle structures like that of the visual system—with many particle structures receiving partial information about the environment (light receptors dedicated to a small region of the visual field, in the case of vision) and integrating this information in successive stages of increasing local integration. In this way, the two-dimensional signal carrying information about the three-dimensional space of the environment is converted to a three-dimensional representation of the environment.

In summary, the two-dimensional nature of the visual signal, stemming from multiple receptors relaying different, but related, information about the environment simultaneously, allows the formation of a three-dimensional representation which produces the distinct qualia of visual experience—the seeing of light patterns and the awareness of objects situated in space. These qualia are highly important in humans and form the major part of our conscious state, unlike in other animals which may rely much more on the sense of smell or the sense of hearing. Those senses derive from signals with a lower dimensionality and therefore must form other kinds of (probably simpler) representations entailing different qualia of the experience associated with such sensory events. We will discuss them briefly in the next subsections.

Hearing

The human auditory system, like the visual system, has multiple stages of processing, starting with the primary auditory cortex and moving on to stages that integrate the signals from the previous stage. Unlike vision, which is based on a two-dimensional signal of light patterns activating multiple receptors simultaneously, hearing is based on a time-varying zero-dimensional signal, i.e., simply the variations of the air pressure at a point in space. The structure of the hearing organ—the cochlea—however, transforms the temporal dimension of the signal into a spatial one. It contains specialized receptor cells, each tuned to a different frequency of air vibrations, thus decomposing the temporal signal into a frequency representation over a short stretch of time (a few milliseconds). Thus, the auditory signal effectively becomes a one-dimensional sensory signal, with the dimension being constituted by the frequency decomposition of the time-varying zero-dimensional signal.

Just like we imagined the visual system in an abstract sense looking as a pyramid made up of successive layers of integration of information, we can picture the auditory system as a geometrical figure with a one-dimensional base, namely, a triangle. The base of the triangle consists of layers with tonotopic organization, i.e., neurons tuned to specific tones, or auditory frequencies. The successive layers integrate information from the lower layers, and the neurons there respond to more complex sound features, such as harmonies (both in frequencies and in time), syllables, etc. One part of those higher layers processes ordinary sounds, while another part is tuned to the specific sound patterns of speech elements and is connected to the language areas in the cortex.

The transformation of the frequency patterns of sound to the internal imagery of conceptual mental states, which is the foundation of language abilities, happens for the most part automatically, without the involvement of conscious awareness. This means that the neural structures performing this transformation have specific, dedicated circuitry able to perform this signal transformation. On the other hand, damage to primary auditory cortex, just like damage to primary visual cortex, makes one unaware of the auditory (or visual) information, suggesting that this part of cortex should also participate in the shared entangled conscious state.

It is a curious phenomenon that people with extensive damage to these areas develop a (usually transient) condition of unawareness of their own deficit, i.e., they become unaware also that they are not able to see or hear anymore. We can think of this phenomenon in terms of our picture of the mental state as a representation of the world with complex structure as an absence of part of the representation, namely, the part that drives the perceptual part of the state. If that part is missing, the faculty of perception is missing too, i.e., the boundary between the dissolving and the constituting part of the mental state is not formed, because the part of cortex where it is supposed to be formed is missing (this is also the part of cortex which gets more active in terms of neural firing, as we discussed earlier, because it is the boundary between the two representational parts). In that case the boundary forms in other areas of cortex and has the meaning of the self perceiving internal mental states, such as when we remember or daydream, which is a perfectly normal mental state not suggesting any striking cognitive deficit. So, the self is complete because the representation of something missing as visual information is formed in the same areas as the representations of something present as visual information, and if this whole area is missing, then the representation of the visual information as a whole is absent from the conscious state. This explains the phenomenon of prosopagnosia, described above

Another important property of vision and hearing, which sets them apart from the other senses, is the veridicality of their representations with respect to the represented objects in the outside world. The close match between the transformation of the object in the environment and the transformation of its respective representation in the brain formed through the senses of vision and hearing is made possible by the high speed of propagation of the visual and auditory signals. For the immediate environment of the organism this speed is effectively instantaneous—any change in the reflectance or vibratory patterns of the object is immediately reflected in the sensory data. (In the case of hearing the speed of sound is slow enough to allow location detection based on tiny differences in the arrival times of sound waves in the two ears.) This nearly instantaneous causal relation allows for the formation of a representation in the brain which transforms in synch with the transformation of the object in the environment and its corresponding sensory data.

Although the light signal is naturally two-dimensional and allows the formation of visual qualia and experience, while the auditory signal is naturally one-dimensional, leading to the formation of auditory qualia and experience, it is possible to form a two-dimensional image based on the pattern of sound waves reflected off an uneven surface. This phenomenon is exploited in the technique of ultrasound imaging, widely used in medicine, and it is also found in nature where some animals use a similar technique for navigation and prey location. Bats have not just one sound wave receptor, as is the case with the human ear, but a whole array of receptors located on their branching antennae. They effectively form a two-dimensional array of receptors, such as the retina of the eye, but with much lower resolution. Since it is much more complex to use the information in the reflected sound pattern to form an image than the visual information on the retina, bats probably do not experience visual qualia from their echolocation, but something intermediate between vision and sound, just enough for knowing where the prey is and to distinguish it from obstacles in their flight path.

The other major senses—of smell, taste, touch and temperature—do not lead to an internal image of the experienced object, due to the nature of their signals, but they have other effects on the conscious mental state which will be explored next.

Other Senses

The senses of smell and taste are based on the detection of chemical substances which interact with specialized receptors to produce neural signals. Those signals evoke the release of neurotransmitters and possibly predefined patterns of neural firing in some areas of the cortex, thus modulating the shared entangled state and effectively its internal structure. In this way, they have an effect on the cognitive processes and our conscious perception, but do not form a representation of the object related to the sensation, as it is the case with vision. The reason for this is the way those signals propagate—the molecules of the chemical compounds are carried by air or the watery environment in the mouth to the receptors, so there is no direct temporal relation between the behavior of the object and the perceived sensation. The signal is also one-dimensional, or, at best, made up of multiple one-dimensional sensory events which are not necessarily temporally related to each other.

The sense of touch also does not provide a continuous stream of information of the behavior of an object in the general case, since we usually come into contact with an object only briefly, and therefore it does not produce a sustained image of the object in the brain. Such relation, however, can be learned through practice, based on the relative high density of the receptors in some parts of the body, such as the finger tips, as is the case with Braille reading.

The skin is also endowed with receptors for temperature, which are very useful for the survival of the organism since temperature is an important factor for the biochemical processes of life. Like smell and taste, the sensations produced by this sensory system only modulate the conscious state but do not form a representation of the object, in this case the air or water in the immediate environment.

The sense of pain is also mediated by special receptors which are present not only in the skin but also in the muscles and other body tissues. It is related to the emotional systems of positive and negative feelings in general and it will be discussed in the next section.

What are the Emotions

The emotions can be divided roughly into two main categories—pleasant and unpleasant, or positive and negative emotions. There is a variety of different basic emotions belonging to each category, such as love, sexual desire, happiness, joy, laughter, etc. in the first one, and pain, anger, sorrow, guilt, fear, etc. in the second one. Each of them has a distinct feel associated with it, or in other words, distinct qualia of experience, and it is also possible to experience them with different intensity and to a different degree, even multiple emotions simultaneously.

Since we assumed that the conscious mental state and its associated qualia of experience are manifestations of the indirect entanglement of particle structures in the brain and their transformations in the metabolic reactions of the life processes in biological organisms, we need to cast the explanation of the phenomenon of emotions in that framework as well. So, we can hypothesize that emotions are some kind of properties of the conscious mental state and its behavior as it transforms with time. The dichotomy of positive vs. negative types of emotions suggests that they must be physically realized by two opposite tendencies in the constitution of the mental state, and we actually have encountered such dichotomous tendencies already—those of coherence and decoherence of the shared entangled state.

Thus, we can suppose that a process of greater representational coherence, leading to a greater degree of reconciliation among the individual representational elements, is experienced as a pleasant sensation, while a process of representational decoherence, leading to less reconciled representational structures, is experienced as a negative sensation. This picture matches also the intuitive perception of pleasure as associated with cognitive agreement and harmony, and displeasure as associated with cognitive disagreement and disharmony.

We also need to imagine the degree of representational coherence not as a static entity, but in terms of the dynamics of the evolution of the conscious state. We said earlier that the conscious state constantly constitutes itself, with one part being constituted and another one being dissolved. The degree of coherence, then, in dynamic terms corresponds to the degree of preparation of the elements of the constituting part, and, as we saw earlier, that involves unconscious processes of particle entanglement outside of the main shared state. So, the emotions we experience are related to the cognitive content of the mental states and their remote relations actualized as unconscious connections.

At the physical level of particle structures the emotions would depend on the properties of the assembly processes of the physical structures into a unit with a shared entangled state. Since these processes are essentially the same as the basic metabolic transformations of biochemical compounds, it follows that the character of the conscious state over a medium stretch of time (e.g., between minutes and a day) is rather stable. In other words, one's personality and basic knowledge usually doesn't change dramatically in such a period of time, only the conscious state traverses the landscape of possible conscious states within the limits of those personality traits or basic knowledge space. In this picture, the emotions represent a transformation of the landscape of the knowledge space due to a transformation of the biochemical processes that determine the inclusion of particle structures in the shared entangled state.

It is no surprise, then, that neurotransmitters and substances that alter the mechanisms of release and uptake of neurotransmitters in the synapses of the neurons can affect the emotional experiences of a person to a great deal. These substances directly affect the metabolic reactions going on inside the cells of the brain, and thus affect the ease with which particles structures from different cells can interact and form an entangled ensemble. Some substances make this process easier, and they have the effect of producing pleasant sensations, like those of the opiates, while others make the process more difficult and they have the effect of reducing the pleasant sensations, like some antipsychotic drugs. Toxic and harmful substances, by their very nature, have a disruptive effect on the metabolic processes in living organisms and they can induce negative sensations directly if they can get inside the brain. The effects on the transmissions of the neural signals naturally alter the structure of the conscious mental state as well, and therefore the content of the train of thought to some degree.

The different emotions, representing different regimes for the operation of the entanglement mechanism and therefore different cognitive landscapes which the mental state can traverse, allow for different reasoning and behavior patterns, thus effectively increasing the capacity of the brain for reasoning and memory. When these regimes are appropriate for specific situations and circumstances in the life of an organism they can increase its success in solving daily problems and lead to better chances of survival and procreation. In this way, emotional reasoning has an evolutionary advantage over plain reasoning without emotional modulation and is one of the major improvements in the course of the evolutionary development of biological organisms.

From modern brain imaging methods we know that emotional states in the brain are controlled by the activity in the limbic system—a network of anatomical structures situated near the centre of the brain, underneath the cortex. These anatomical structures are also called the "reptilian brain" because they first appear in the evolutionary record in the brains of reptiles. Fish don't have a limbic system and their behavior is simpler than that of reptiles, while mammals have in addition a large cortex and their behavior is more advanced that that of the reptiles.

The structures in the limbic system send axonal projections to all parts of cortex, allowing them to control the mechanisms of release and uptake of neurotransmitters as well as to send signals and thus participate in the shared conscious state. In this way, apart from modulating the ease of transmission of neural signals through the neurotransmitters, the limbic organs can also participate directly in the representations formed by entangled particle structures. Since they are much smaller in size than cortex, they are unlikely to affect the content of the mental states, which depend on the specific configuration of particle structures participating in the shared entangled state, but they may be capable of modulating the degree of coherence or decoherence of some or all of the particle structures, thus affecting its trajectory in state space.

When only part of the shared state's coherence is modulated, then the emotional state has a cognitive component resulting in emotional reasoning processes like compassion, remorse, etc. In those states, the coherence among and within the different parts of the mental state is altered with respect to normal, non-emotional reasoning, and consequently the reasoning process is tinged with the specific feel of the respective emotion and it follows a different trajectory leading to different beliefs and conclusions compared to the non-emotional reasoning process. The activity of the limbic system orchestrates the emotional reasoning process and when it remains inactive, i.e., still exchanging signals with the cortex, but only representationally compatible ones which do not modulate the activity in the cortex, then the reasoning process is non-emotional.

Under normal circumstances the mental state always has a cognitive component and the distinctly recognizable emotional state comes in addition to that component due to modulatory effects of altered synaptic transmission mechanisms and representational neural signals. So, the emotions are deviations from the normal functioning of the brain, which can be supported for relatively brief periods of time since they represent altered metabolic activity and such activity cannot be sustained indefinitely. If that happens, nonetheless, then the brain enters an abnormal regime of functioning which we recognize as a mental illness and which will be discussed in the next section.

Besides the modulation of the coherence of the components of the mental state there is also one other logical possibility of the effects of the external signals from the limbic system on the shared entangled state in the cortex, namely, a pure disruptive effect on the representational coherence which can be described also as injection of noise into the representational structure. Such an effect is easy to realize, since it does not require a structured signal, but just an incompatible signal with respect to the representational coherence of the mental state. This kind of signal would be experienced as the exact opposite of the smooth flow of well-prepared cognitive states, i.e., as painful sensations.

The perception of pain is mediated by dedicated receptors found in most parts of the body (but not in the brain) which send signals to specific areas in the brain. It is possible to alleviate or eliminate completely the sensation of pain by blocking those signals or by infusing substances (like opiates) in the brain which facilitate the transmission of signals among neurons and thus increase the degree of coherence in the mental state. In both cases the disruptive signal is either masked or eliminated completely, but the greater degree of coherence induced by opiates has the side effect of altering the internal structure of the mental state and its state space, i.e., it produces an altered state of mind.

What are Mental Illnesses

An illness is a deviation form the normal, healthy functioning of the human body and any part of the body can be subject to such a deviation, including the brain. When the brain starts functioning in abnormal ways this affects not only the chains of metabolic reactions, but also the mental processes supervenient on those reactions, leading to psychiatric, or mental, disorders.

Before discussing the mental illnesses specifically, we will discuss the concept of health in general and its specific peculiarities. Health, as we said, is a state of normal functioning of the metabolic reactions, meaning that the substances making up the body are in the right quantities and in the right places. The biochemical reactions are balanced in the sense that they have the right volumes of inputs and produce the right volumes of outputs. In terms of representations we can also say that the representations formed by the particle structures throughout the entire body are in balance, i.e., reconciled to a high degree and correctly reflecting each other.

This state of balance is a steady state in the terminology of dynamic systems theory and represents a local minimum in the landscape of all possible states. However, since the body is a very complex dynamical system, the landscape contains other local minima not far away and the body could switch to a different regime of functioning if pushed into another local minimum nearby by external forces. In that case, the balance of the metabolic reactions will be different and there is a high chance that some of the inputs or outputs of a reaction will be in over—or undersupply. If the body continues to function in this regime of over—or undersupply of a biochemical compound, it would gradually accumulate substances for which there is no use in the body, or it will run out of needed substances, and some of the biochemical reactions in the complex metabolic network may halt altogether. This can be very harmful to the body and could lead to cell death, death of whole organs, and the death of the entire organism.

In terms of representational structures, this process of disbalancing and gradual degradation of the organism can be viewed as a process of accumulation of errors in the representations, i.e., accumulation of negative representations, ones that do not fit in the representational complex of the body. A typical example of such a process is a foreign organism invading the body (e.g., a virus, bacteria, or even some macroscopic organisms like parasitic worms living in the intestine). The foreign organism is a self-sustained representation which sustains its own existence and reproduction, but does not sustain the functioning of the host representation, and if it manages to alter the host representation more than the host representation alters it, it can lead to the complete disruption of the host representation, i.e., to the death of the host organism.

Since the biochemical processes in organisms of the same kind (i.e., the same species) are for the most part identical, or very similar, the ways in which the complex network of metabolic reactions can be disbalanced are also very similar for different individuals belonging to the same species. These different ways of disbalancing is what we call diseases, and there are also typical ways of rebalancing the biochemical functioning, and these are the standard treatments that lead to the curing of the diseases. However, since every individual is slightly different from any other one (and in the vast numbers of particles making up an organism and biochemical reactions that these particles engage into there is a lot of room for tiny differences), the degrees of the deficiencies and the responses to the standard treatment are all slightly different as well, and in some rare circumstances the standard treatment may fail to achieve the rebalancing of the biochemical processes.

All these peculiarities are reflected in the nature of our health care systems. When our biochemical processes get disbalanced (which typically evokes pain sensations, because pain is just a signal signifying incompatible representations), we go to the doctor who sets a diagnose, i.e., determines which specific kind of disbalance occurred in our bodies, and prescribes a treatment, most often in the form of a medication that we need to take, i.e., an external substance which will rebalance the quantities of the substances in the body or induce such a rebalancing by altering the metabolic mechanisms at work in the body. Most often this works, but in some circumstances the treatment needs to be readjusted after some time or it may even fail to work at all and the disease stubbornly persists.

The disbalancing of the metabolic reactions in the brain, as we said, has the side effect of disbalancing also the thought process and the belief systems and reasoning patterns of an individual. This usually happens at the level of biochemical substances and their actions as neurotransmitters in the mechanisms of signaling among neurons. Therefore, most psychiatric disorders can be affected to some degree by medications that rebalance the levels of these neurotransmitters, as is the case with treatments of schizophrenia, depression, phobias, etc. These medications typically affect also the sleep patterns of the patient, since sleep is also a different regime of functioning of the brain.

The case of schizophrenia is a very curious one, since the disease has a well established symptomatic of the type of delusional thoughts that patients engage into and which the pharmacological intervention can reliably reverse. It seems as if taking a pill magically erases specific beliefs and implants other ones, as in science fiction stories. However, this seemingly magical action of the medicine can be explained in terms of rebalancing the substances of the brain metabolism, which, when disbalanced, push the thought process in a specific direction leading to a specific kind of false beliefs.

The typical stories told by schizophrenic patients involve perceptions of being spied upon, or other people being able to read the patient's thoughts and vice versa. They involve also beliefs of miraculous acts and events happening around them, usually attributed to aliens or supernatural and religious agents. Such beliefs are based on actual visual and auditory hallucinations experienced by the patients, which resemble dream phenomenology mixing up with the awake sensory experience. It is as if there are some other beings inside the head of the patient, who alter the patient's own experience of the external reality, and implant foreign thoughts and experiences on him or her. That is the reason why the disease is called schizophrenia, meaning literally 'split personality'.

The idea of the split personality is actually the key to understanding this disease in terms of our picture of the mental state as an entity in abstract state space realized by shared entangled particle states. We considered the self as being realized by the interface of the two parts of the mental state—the dissolving part and the constituting part. This is also where perception occurs when the representation of the self at the boundary of the two parts interacts with the representation of the environment derived from the sensory organs activity. In order for the self to experience hallucinations, i.e., perceptions that do not correspond to events in the environment, the transformational process from the original sensory activity to the representation interacting with the self must get confused, i.e., the transformation must not be veridical, but it must form its own representations. In other words, the representational transformation from the sensory organs to the self in the healthy brain is linear, i.e., it transforms in stages from the original activity to the final representation without being influenced substantially from other activity, e.g., memories. In the schizophrenic brain the transformation is not linear, but self-reflective, i.e., it intertwines with other internal representations in the brain of memories and conceptual states, forming an entity in its own right which amounts to a rudimentary conscious mental state. Thus, it is indeed the case that a second personality, or mental state, emerges in the head of the patient, albeit a weaker and not entirely conscious one.

This second mental state (or maybe multiple small ones) follows its own logic and evolutionary trajectory, interacting occasionally with the main mental state of the person, which is experienced by the schizophrenic as an external voice or a vision of something which is not there. It also explains why the schizophrenic patients believe that their hallucinations are acts of supernatural beings or aliens—this is the only way to explain the feeling of an external entity (in this case the second mental state) affecting one's own thought process and perceptions without any apparent cause coming from the external environment. This also feels like mind reading abilities of other people whom the patient encounters. In the patients interactions with them he or she feels that their behavior and personalities are full of surprises, due to the interferences of the second mental state in the perceptual process, and that is interpreted as supernatural abilities like mind reading (supported also by the continuously present feeling of something else manipulating the patient's thoughts). In the somewhat milder cases of the disease the perceived odd behavior is interpreted a bit more rationally as hidden intentions, such as those of spies.

The hypothesis of a second centre of mental activity emerging in the schizophrenic brain is supported also by the nature of the physiological changes found to be present in most of the schizophrenic brains, namely the increased effectiveness of the dopaminergic signal transmission between neurons. The typical antipsychotic drugs act by blocking certain dopamine receptors which are part of the neurotransmission system between the neurons, thus reducing the effectiveness of dopamine as a neurotransmitter. Dopamine is one of the main known excitatory neurotransmitters, and its enhanced function is associated with positive feelings of happiness and well-being. By our hypothesis regarding the nature of emotions and the properties of the mental state, enhanced dopamine effectiveness would lead to a more concentrated mental state and also more reconciled representational structures, meaning also more positive emotions. This more concentrated mental state, when imagined as a particle-like entity in abstract state space, would leave more empty room in its immediate environment, allowing for a second entity of the same type to appear there. There are probably many other preconditions that need to be met for this to happen, but the higher concentration of the abnormal mental state seems to be of central importance for a personality split.

Besides the higher concentration of the mental state due to enhanced effectiveness of the excitatory signals among neurons, the brain physiology can be disbalanced also in the other direction—towards a more diffused and weakly coupled mental state. The feelings associated with this condition are likely to be ones of tiredness, lack of desires, unwillingness to act and hopelessness regarding one's future. These are all typical symptoms of a depression. So, depression is in a sense the opposite of schizophrenia, i.e., a decreased efficiency of the excitatory signals leading to impeded transitions of the mental state from one well-defined state to another. Such a regime of functioning can be induced also by higher doses of the antipsychotic medications, but there are also many other possibilities for reducing the excitatory coupling of neurons besides the suppression of the effectiveness of the dopaminergic signaling system.

In general, we can say that all psychiatric diseases represent abnormal regimes of functioning of the physiological processes in the brain and are based on disbalances in the presence of one or more substances participating in the complex chains of metabolic reactions constituting the activity of the healthy brain. These abnormal regimes can (and normally must) change the typical trajectories of the mental state, thus altering the person's beliefs and rationalizations of the events around him or her (what is called 'madness' in vernacular speech), but the opposite is not true, i.e., one's thoughts cannot affect the disbalances and the biochemical processes at the molecular level, except sometimes indirectly in the most extraordinary set of circumstances.

Aging of the Brain
(Developments Throughout the Course of Life)

When we talk about the brain in general we usually have in mind the healthy adult brain, but in order to have a complete picture of this very special organ of the human body we need also to have understanding of its life trajectory, form the moment of conception till the death of the organism, and the major transformations it undergoes in this time frame. If we were to represent the life trajectory by a graph depicting some measure related to the brain, like size, performance, complexity of organization, memory capacity, etc., most of those graphs would look like an inverted U-shape curve, going up in the initial stages of life—childhood and adolescence, peaking somewhere during the adolescence or the middle ages, and slowly declining afterwards until the death of the individual. This is basically how we should think of the brain in developmental terms—as an organ which grows, matures and declines, just like the body as a whole.

Earlier we imagined the body of any biological organism as a concentration of representations physically realized by intertwined structures of particles and processes mutually adapted to each other. We also hypothesized that the representations start accumulating at the moment of conception of the organism, leading to a dichotomous process of accumulation and dissipation of adaptations. Due to the annexation of new particle structures to the organism during the period of growth, the processes of accumulation and dissipation are in balance and the organism grows in complexity and representational richness, but once the physical growth stops, the dissipative processes prevail and the decline of the organism begins.

This picture reflects also the life trajectory of the brain, as the organ of the body with the highest concentration of representations. Until adolescence the brain grows in size, albeit at a slower rate than the rest of the body, since it starts with a relatively larger size at the moment of birth compared to the rest of the body. During this period of growth the concentration of representations in the brain is extremely high, physically manifested by a higher number of synaptic connections among the neurons, with the result of making the entanglement of the states of the particle structures in the brain unusually strong and therefore the mental state highly concentrated. This is the reason why children do not have the full intellectual skills of an adult—their mental states have less internal structure because of the high concentration of the mental state, but the state is more agile and more rapidly transitioning, producing the lively behavior of children.

In late adolescence the brain undergoes a massive pruning of synaptic connections, which has the effect of slowing down the mental state, or reducing its "energy", in some abstract sense, and this allows the more complex and richer thought process of the young adult individual. The pruning of the connections, cell death and loss of flexibility in the life processes of the neurons continue at a slow rate after adolescence, leading to a gradual degradation of the cognitive abilities and the richness of the thought process of the adult. During old age the effects of these processes are highly visible—the brain has lost nearly half of its mass and has shrunk in size, learning new facts becomes much more difficult, and the chances of a mental dysfunction increase.

The other side of the dissipation of representations throughout adulthood is the possibility of accumulation of knowledge. As the entanglement among particle structure relaxes and the mental state losses "energy", the configurations of particle structures become richer and acquire finer detail, making the mental states more complex and therefore the knowledge of the individual richer. This process continues until late adulthood, with knowledge and experience peaking late in life, just before the onset of the old age, unlike most of the basic cognitive abilities, which peak earlier. In terms of representations, we can say that the most complex representations, corresponding to most profound and rich knowledge, occur at the moment of the most relaxed particle structure relations, while there are still enough particle structures in the brain that engage in the mental state. From that moment on the relaxation continues, but the dissipation of representations in the form of disengagement of particle structures from the shared entangled state continues at a higher rate, resulting in a net decrease of complexity and richness of knowledge content.

In an abstract sense we can say that the early situation of strong entanglement of more flexible particle structures is also a state of more open possibilities for how those structures will relax and form conceptual content, while the later situation of more conceptual structure realized by weaker entanglement of more diverse and less flexible particle structures is a state of fewer open possibilities and more actualized possibilities, constituting a larger volume of knowledge.

The gradual entrenchment of conceptual mental states in the course of life has manifold effects on the cognitive abilities and the belief system of the individual. We can say that the imagination of a person gradually declines in its expressiveness, from the highly productive and uncontrollable imagination of a child to a more subdued and less creative imagination of an adult to a nearly complete lack of imagination in the later stages of life, which are dominated by memories of the past, i.e., well entrenched mental states. A similar development trajectory characterizes emotions as well. The emotional experiences of a child are very vivid and rich in nuances, which gradually become duller and less frequent in adulthood, giving way to rational reasoning. This fits with the picture of emotions as relations of representational agreement, or reconciliation, among the different components of the mental state. The stronger coupling of particle structures in the earlier stages of life allows for stronger interactions between the constituent parts of the representation, while the relaxing of the coupling in later stages of life and the entrenchment of the conceptual states reduce or eliminate the phases of representational disagreement in the dynamics of the evolution of the mental state.

In general, we can say that the brain undergoes a process of massive accumulation of representations (in terms of adapted particle structures) during its growth phase, which encompasses the first few years of life, and gradual dissipation of representations afterwards, resulting eventually in a breakdown of some of the basic mechanisms supporting life and the death of the organism. The initial accumulation of representations has two main sources for the accumulation process—the internal genetic plan manifested through the processes of cell division and cell specialization, and the external influences from the environment, initially in the form of biochemical signals from the mother's body and later on as sensory inputs and biochemical influences affecting the body through food and other substances taken in from outside.

This figurative division of the sources of representational influences on the developing brain (and body in general) prompted a scientific debate regarding the role of 'nature' vs. 'nurture' in the development of the organism and also the mind of a human being. The positions taken initially were more radically in favor of one or the other extreme, but eventually the scientific community began to appreciate the complexity of the developmental process and the significant role each component in this process plays, so the debate has transformed from a dichotomous "either-or" set of arguments to questions about the exact role of each internal or external factor in the developmental process.

Still, we can say in a very general sense that the balance between the internal (genetic and biochemical) factors and the external (sensory and biochemical) factors is initially tipped heavily towards the internal ones and gradually shifts towards the external ones. This transformation is not always smooth and gradual, but has sharp phase transitions and even structural rearrangements, but overall the trend is towards an increasing influence of the external factors.

In a more abstract sense this property of the development of the organism can be defined as an interactional relation between the representational complex of the organism and the representational influences from the environment. Using our picture of representations as particles in some abstract higher-dimensional state space, this process would look like a solid object composed of many particles (this is the representational complex of the organism) being bombarded by particles from the environment, which are less dense, (these are the influences from the environment) causing it to gradually relax and dissolve while affecting and transforming the environment with its own representations in the process of its degradation.

Within the solid structure of a mature organism we can picture an inception of another structure, starting initially with a much higher density of representations and eventually splitting off the original structure and taking a life trajectory of its own. This is the process of reproduction of the organisms, sparked by an exchange of representational information between two mature (and partially relaxed) structures.

This picture of representational structures in abstract state space is informative not only about the aging process of biological organisms, but also about the nature and role of life in the universe. In representational terms, life is a process of accumulation of more representations resulting in more complex structures of simple representations, and therefore aging should be understood along those lines as well—as a relation between the representational information of the organism and the representational information of the environment (including other organisms). The exchanges of information may go either way, but the net effect should be a net increase in representational information (quantifiable in some way), in which a major role plays the faculty of human consciousness and more specifically the ability to learn by forming new conceptual structures.

After this sketchy overview of the main cognitive faculties pertinent to the functioning of the brain which unfolded in the preceding sections of this chapter, we can tackle the more fundamental question of free will, as formulated in philosophy, and eventually build a picture of everything that exists, i.e., what we call 'the universe'. This is now possible because, according to our philosophical understanding of the world, there are only two fundamental categories of 'stuff' that exist—matter and mind. We built our understanding of the former in the first chapter, and the discussion in the second chapter defined the latter in similar terms, as the same kind of 'stuff' as our conception of matter derived from the first chapter. Thus, the single framework for understanding both matter and consciousness as one and the same fundamental entity will allow us also to build a picture of everything that exists, and also provide hints to the answers of the questions of why it exists and what could be its purpose and future development. These issues will be discussed in the next chapter.

Is There Free Will

One of the most difficult and long-standing questions in the philosophy of mind concerns the apparent ability of humans to act with deliberate intent, denoted in the terminology of the philosophy of mind as the faculty of 'free will'. In fact, the idea of free will is rooted in the dualistic doctrine within philosophy of mind which derives from Christian religious beliefs that were dominant during the times of the Renaissance in Western Europe. Modern day scientists have no trouble seeing the human brain as a mechanism, in purely materialistic terms, and consequently the human mind as a mere epiphenomenon arising from the communication among neurons via the mechanism of neural firing. From that perspective free will seems more like an illusion and there have been attempts at supporting this claim with proper empirical evidence from cognitive psychology experiments.

Most notably, some researchers were able to demonstrate that the EEG signal from certain areas in the brain shows a consistent rise a few hundred milliseconds prior to a voluntary movement produced randomly by the volunteers in the experiment and in another experimental paradigm that the participants can be tricked into believing that an action on their side causes an event on the computer screen when in fact there is no causal relation and vice versa, that their actions are unrelated to the events on the computer screen, when in fact they were causing them. All this is taken as supporting evidence that our deep seated conviction that we act according to our free will might be an illusion.

Such arguments notwithstanding, the question of free will is closely related to that of determinism, namely, whether the movements of matter in the universe are deterministic, i.e., completely determined by their prior states, or whether there is an element of novelty emerging at every time point, which would render them non-deterministic. If the universe is deterministic, then the human brains as parts of it would also be deterministic mechanisms and our conception of free will would turn out to be an illusion, and if not, then the creation of novelty in the interactions among physical particles could be a source of unpredictable novelty and uniqueness in our macroscopic behavior and would fit our intuitive notion of free will.

According to this definition of free will we have to answer the question regarding its existence positively within our framework of physical matter as representations, since, as we saw in our discussion on what is life and how it leads to consciousness, the process of constant accumulation of representations results in the constant creation of new representations, which is exactly the element of novelty that we consider as constituting the free will of a conscious being. On the other hand, the processes of interactions of the representations and the emergence of novel representations are governed by fixed physical laws and in that sense the evolution of the vast collection of representations that constitutes the universe is deterministic (which, however, is different from knowable), seemingly contradicting our earlier conclusion. The contradiction would disappear, however, if we observe that deterministic might be different from knowable—even in principle—and that the unknowable parts of the representations and their interactions is what would naturally constitute the elements of novelty and consequently the free will of a material body within the universe.

Thus, in order to answer the question of free will we need to consider what is knowable and what is not, and the best way to approach this question is to turn again to the idea of representation as underlying the concept of knowable as a matter of principle. To know something, in the general sense of this word, means to have a mental state of understanding something about the world, in other words, the knowledge encoded in the mental state has to correspond in some way to the state of affairs in the world in a veridical way. Since we regard all matter as representational in nature, the mental states are also representations (with more complex content than the material particles that make them up), and in this way what is knowable turns out to be equivalent to what is representable in the human brain.

Now, since the human brain is a result of an evolutionary process of accumulation of representations, it can represent only those aspects of the world that played a role in the evolutionary process, namely, it can form mental state representations of macroscopic objects, their movements and their relationships to each other, but it cannot form representations of, e.g., the microscopic states of the objects or the internal workings of the brain. In general, in a closed system of representations only a subset of all representations can in principle be represented by other representations, and there is always a subset which only represents, but is not represented itself—these are the most complex and most advanced representations at the forefront of the process of accumulation of representations. This includes the newly emerging representations in the form of new human knowledge as a result of the constant accumulation and compression of the representations in the human brain.

Since in an evolving universe there is always a subset of representations which are not represented by some other ones, it means that there is always a subset of objects whose inner workings is unknown. Although they can be regarded as deterministic mechanisms in principle, it is also not possible in principle to know and predict the behavior of all of them at the same time.

For example, let's assume that our science will develop so much in the future that we will be able to build a computer model (possibly involving quantum computers) which would adequately simulate the behavior and inner mechanisms of a simple insect, e.g., a bee. This means that the bee can be understood as a mechanism, in a way we understand a car or a computer, and we would be reluctant to attribute free will to it. Note that we would still be unable to predict the behavior of a real bee using that model, since there is always a degree of indeterminism in the interactions among particles at the quantum level, but we will be able to predict the average behaviors of large number of bees. Similarly, we cannot predict the exact behavior of the particles making up any other mechanism, such as a car or a computer, and consequently we cannot predict when it will break down, for example, but we still deem such mechanisms as lacking free will.

In order to acquire the knowledge of the inner workings of a bee, however, we would need to develop a lot of other knowledge and complex instrumentation, which in turn will make the patterns of particle interactions in our brains even more complex, i.e., it would result in the formation of even more complex representations elsewhere, namely, in the human brain. Also, it would allow us to develop technologies that connect with the brain and augment its capabilities, so by the time we become capable of understanding the ordinary human brain, most people would have their brains and bodies enhanced so much that they would form a complex conscious entity of its own with even more complex internal states. So, again, the goalposts will be moved and there would be a subset of representations in these more advanced brains which are unknown and unrepresented anywhere else.

In this way, we can conclude that although the interactions of matter in the universe might be governed completely by physical laws and therefore be entirely deterministic there still will be facts about the material objects that remain unknowable, even in principle. A key consideration in this argument is that the universe evolves in a process of accumulation of representations and the representations are not arranged in some special way by some supernatural force or being. The unknowable part of the representations is what would naturally correspond to our intuition of free will—it can affect the other representations, but it does not get affected by them to a matching degree and therefore it can determine completely the fate of the other material objects but they cannot determine its fate completely. This is pretty much the relation between humans and the other animate and inanimate objects they interact with (almost exclusively on the surface of the Earth), which can be conceptualized also as humans imposing their free will on the rest of the world.

The Space of Mental States

Finally, in preparation for taking on the question of what is the universe, we need to clarify our picture of the abstract space formed by the conceptual mental states that constitute the totality of human knowledge and experience. Earlier we noted that this abstract state space should be high-dimensional, with each concept forming its own dimension in the general case, since each concept can exist independently of the other ones and its content can vary. We need to note right away, in addition, that those dimensions would not extend over long stretches of space, as is the case with real, three-dimensional physical space, but these would be 'tiny' dimensions existing mainly near the origin of the coordinate system, where the concept is well-defined. Their small extent is due to the small variation in the content of a concept.

The transformation of a person's mental state, a.k.a., the 'train of thought', would trace a trajectory in that space just like the movement of a particle in real physical space. This movement, however, would represent a process of jumping from one dimension to another following the sequence of concepts activated in the train of thought, with some jitter from the excursions of the state in several dimension at the same time corresponding to co-activations of several concept in a single conceptual scheme. When those excursions get deeper and more frequent in some locations they would be likely to engender the birth of a new dimension, i.e., the formation of a new concept due to accumulation of knowledge which cannot be expressed in terms of the existing conceptual structures. This mechanism ensures that the dimensions remain tiny, since pressures on extending the state space result in the birth of new dimensions, rather than the extension of the existing ones.

This picture is radically different from the behavior of matter in real physical space, where the dimensions stretch without limits in space and the particles can move in any direction indefinitely as long as their movement is not disturbed by interactions with other particles. However, apart from this fundamental difference, there are many commonalities between the abstract state space and physical space.

One of them is the compatibility and incompatibility of certain pairs of states. For example, the concepts of 'black' and 'round' can be coactive at the same time during a perceptual process of, e.g., perceiving a black snooker ball, meaning that the mental state in abstract state space occupies the two dimensions of blackness and roundness at the same time. The concepts of 'white' and 'round' can be coactivated as well, but this is not the case with the concepts of 'black' and 'white' (or any other color) during the perceptual event of a snooker ball. This means that the mental state cannot be occupying such pairs of dimensions at the same time, very much like some of the limitations on the behavior of physical particles, which cannot be in a spin-up and a spin-down collapsed state simultaneously, for example.

Another common feature between the behavior of particles in physical space and in abstract state space is the reproducibility of the behavior in identical (or very similar) circumstances. Human reasoning consists largely of reproduced patterns of thought and only rarely do we come up with a novel idea which we consider an insight, allowing us to learn something new and increase our knowledge. This phenomenology parallels closely that of the biochemical reactions in living organisms, where the majority of the molecular transformations are reproductions of well-established metabolic reactions and only rarely the metabolism takes a new path and even more rarely this new type of reaction gets incorporated in the established set of reactions and becomes reproducible. In this way, most of the events involving particles both in real physical space and in the abstract state space of mental states follow well-established patterns and are reproductions of other events of the same type (this is true also for the events happening in places other than the surface of the Earth—most of them are also particle transformations of the same kind, typical for the physical conditions at the given place in the universe).

Still another similarity (which, however, is not perfectly analogous in the small details) is the ability of the particles to participate in nested hierarchical structures. In the mental realm the concepts can be logically instances of other, more general categories, e.g., animal > bird > sparrow, and in the physical realm particles can participate in larger particle structures—molecules > atoms > elementary particles, which behave as units in interactions with other particle structures and therefore can be ascribed distinct chemical and physical properties. We still do not know exactly how mental concepts are encoded at the level of atomic and molecular structures (i.e., what is the physical instantiation of 'animal' or 'bird' in the brain), but if we assume that the particle structures engaged in the shared entangled state of a more general concept include partially (and overlap with) the particle structures instantiating the individual, more specific concepts making up the more general category, then the situation in the mental and the physical realms would be closely analogous.

So, in summary, we can conclude that the physics of real, three-dimensional space and the particles of matter situated in it is closely analogous to the physics of the abstract mental space, where the role of particles is played by mental concepts, in the general phenomenology of the particles' behavior and interactions. The major distinction, which also translates into distinctions in the fine details of how events unfold, between the two realms is the fact that particles in real physical space are very similar to each other and effectively form a space with dimensions with indefinite extensions, while all concepts in mental space are distinct and differ in some respect, making the abstract state space effectively high-dimensional and each dimension with only tiny extension around its corresponding particle-concept.

After this exposition of the picture of the mental and physical space we can now turn to the question of what the picture of the totality of all that exists looks like. Since the totality of everything includes both matter and consciousness (the two basic philosophical categories of what exists in the world), it would be natural to expect that the picture of everything, i.e., of the universe, would be some kind of synthesis of the pictures of the two realms—that of matter described in Chapter 1 and that of the brain and consciousness described in Chapter 2. In other words, it would need to provide a single framework for understanding both matter and consciousness as the same kind of 'stuff', which entails that it needs to combine the two distinct pictures of real physical space and abstract state space in a single unified theoretical framework. So, we would need to understand physical space and state space as the same kind of space, possibly in different stages of its existence (or different phases), but part of the same grand evolutionary process that governs the transformations of the universe. This is the kind of understanding that we will try to develop in the next chapter.

Chapter 3: What is the Universe

Basic Cosmology

The universe is the totality of everything and that includes the two fundamental types of 'stuff' according to the philosophical understanding of the world—matter and consciousness. According to our modern-day scientific theories matter is pretty much everything that exists and it is situated in space and transforming and interacting with the passage of time. Consciousness is a special kind of organization of matter which can greatly influence its transformations, but is not material itself according to the strong materialistic view of the world adopted by most scientists. Since, as far as we know, only a vanishingly tiny fraction of all matter is organized in living and conscious structures, this special kind of organization is considered insignificant for our understanding of the universe, and the discipline of cosmology, which is preoccupied with the study of the universe as a whole, is concerned only with the laws of physics affecting the behavior of matter but not with the effects (and also possible future effects) of consciousness on the transformations of matter in the entire universe.

The reason for ignoring consciousness is that currently it is really of no consequence to the evolutionary process of the universe as a whole. As far as we know, life exists only on the surface of the Earth, and although this seems like a pretty big place to us, it is an extremely minuscule part of the vast expanses of the observable part of the universe. To give some idea of this difference in size, the diameter of the observable universe is about 20 orders of magnitude (10^{20} times) bigger than the diameter of the earth. The diameter of the Earth is about 10^8 times bigger than the human brain, and the brain is about 10^{14} times bigger than a particle, such as a proton or an electron. This makes the brain (and the hypothetical shared entangled states in the brain that we regarded as the physical substrate of conscious mental states) situated near the one-third mark of the scale, about twice smaller than the universe as it is bigger than a subatomic particle.

The difference in scale between a subatomic particle and the observable universe is thus about 42 orders of magnitude (10^{42} times). The distribution of matter in the universe is roughly homogeneous at scales larger than about 10^{21} kilometers, meaning that at scales between 1000 and 10,000 times smaller than the whole observable universe matter starts to clump up and its distribution in space becomes uneven. This continues all the way down to the subatomic particle scale, where the diameter of the particles is about 100,000 times smaller than the diameter of the atoms they make up.

The uneven distribution of matter in space represents the current state of affairs, but it wasn't always the case, according to our current cosmological theories. A few decades ago astronomers found out that the universe is expanding and hypothesized that if we run the currently observed process of expansion in reverse, we must conclude that the universe should shrink to a size smaller than a subatomic particle. This state is called a singularity and it represents the hypothesized origin of the universe and all matter in it. From that initial state the universe started expanding about 13.7 billion years ago and continues expanding today at an even higher rate than before.

So, in the initial fractions of a second after the beginning of time in the singularity state (a.k.a. the Big Bang) the universe expanded at an immense rate, much faster than at any other time, producing a nearly perfectly smooth and uniform distribution of its content, which at that time was just radiation—no particles with mass of any sort—and this is the reason why the universe is roughly homogeneous at the very large scales of distances over 10^{21} kilometers. This scenario is called the inflationary hypothesis and is the generally accepted explanation of the observed distribution of matter in the universe and other phenomena like the variations in the cosmic background radiation, although there are also alternative theories.

After this initial inflationary period, which lasted from about 10^{-36} seconds to 10^{-32} seconds after the Big Bang, the universe continued to expand at a slower rate. During the first second it was filled with immensely hot and energetic radiation, and after that mass-possessing particles started to get created out of the extremely high energies of the massless radiation particles. During the first 10 seconds after the Big Bang particles and anti-particles got created and annihilated until the temperature cooled sufficiently so that these reactions ceased, leaving a small residue of unannihilated particles (protons, neutrons and electrons).

In the following 380,000 years the universe was filled with a mix of roughly uniformly distributed particles which were frequently exchanging photons. In the initial 20 minutes of this period atomic nuclei got formed by the merging of protons and neutrons, and near the end of the period the universe cooled down sufficiently so that the nuclei could capture electrons and start forming atoms. One of the consequences of the phase transition to atom formation was that photons were no longer captured by the charged particles and were able to travel much longer distances. In effect, this meant that the universe became transparent to electromagnetic radiation, and the cosmic background radiation observed today is the consequence of this phase transition.

After the particles started forming, the distribution of matter in the universe started to get more and more clumpy. Since the number of particles remained roughly constant after the temperature dropped down enough so that the formation of new particles out of radiation ceased, the density of matter in the universe continuously decreased due to the expansion of space in which matter was situated. The effect of the force of gravity experienced by all particles is to keep them close to each other, so they started forming clumps of denser matter separated by regions of more diluted matter.

Eventually, about 400 million years after the Big Bang, some clumps of matter, which at that time consisted of mostly hydrogen and smaller amounts of helium and lithium, got dense enough to pressure the protons and neutrons in the centre of the clump to interact via the strong force and merge in heavier nuclei, and so the first stars were born. The stars are the places where all other chemical elements, apart from those three created in the first moments of the Big Bang, get created, and the other consequence of their activity is that the radiation emitted by a star expels all particles in its vicinity outwards, effectively clearing up its surroundings.

In that way, due to both the expansion of space and the radiation coming from the stars, the distribution of matter in the universe gradually became very uneven at the smaller scales, being either empty space or a cosmic body (a star or a planet) of densely packed particles.

The stars group together in larger clumps, called galaxies, which are separated by even larger expanses of empty space, and the galaxies themselves form clusters on a larger scale and superclusters on an even larger scale. So, the degree of "clumpiness" is roughly uniform at all scales where the distribution of matter is not homogeneous, suggesting that the universe was highly homogeneous in the initial stages of its existence with only tiny variations due to the quantum nature of matter, which are indispensable according to quantum theory.

Although at the time when particles were created the entire space in the universe was filled with them, due to the continuous expansion of space their current density in the entire observable universe is estimated at about a few particles per cubic meter on average. Of course, in the interstellar and intergalactic deep space their density is much smaller and in the space occupied by a cosmic body, where they are closely packed, their densities are enormous, so the average number of a few particles per cubic meter gives an idea of the vast expanses of empty space separating the cosmic bodies.

Besides matter, the universe is also filled with radiation, mostly due to the cosmic microwave background. Its density is estimated at about a billion photons per cubic meter—much higher than the average particle density, but on the other hand containing only a negligible amount of the energy content in the universe. Observable matter too forms a very tiny fraction of all matter in the universe, about 4% in total, most of it in the form of individual particles of the lighter elements (hydrogen and helium) in deep space, and only a fraction of a percent packed tightly in stars and planets. The vast majority of the material content of the universe, though, is unobservable with the means of modern science and deeply mysterious. About 23% of it is estimated to be what is called 'dark matter', which is subject to the force of gravity and therefore counteracts the expansion of the universe. The rest 73% is the so called 'dark energy', which is introduced in the theoretical models in order to explain the observed accelerating expansion of the universe and acts in a way opposite to gravity.

So, the most striking feature of this picture of the universe delivered to us by modern cosmology is probably its unfathomable vastness compared to the units of measure we are dealing with in our daily life. We call our environment the 'macroscopic world' and we find it difficult to comprehend the huge numbers of particles making up every macroscopic object, including our own bodies and brains, but it turns out that the other end of the scale—towards larger and larger material structures—is about twice more distant, making our 'macroscopic world' look disappearingly tiny and insignificant.

In the following sections we will try to interpret this picture of the universe in the light of our understanding of matter and consciousness developed in the preceding two chapters, and we will try to explain why the universe is the way we see it in the picture painted by modern cosmology. Namely, why was there a Big Bang, why are there three dimensions of space, why the distribution of matter is so uneven, and ultimately, where does this vastness of all astronomical numbers come from? Understanding what the universe is and knowing its past ought to enable us to predict where its future evolution will lead, so we will also try to construct a timeline of the future and answer the more fundamental question on whether it will have an end or continue existing indefinitely, and in what form.

What is the Big Bang

What we know for sure based on astronomical observations is that the universe is expanding. The main line of evidence comes from the observation of a red shift in the light emitted by distant objects—the more distant they are, the stronger is the shift of their spectral lines towards the red end of the spectrum. From this fact we conclude that since the universe is expanding at the present moment, it must have been smaller and smaller further back in time, reaching a limit of a minimum size, called a singularity, which we consider the origin of space and time.

In terms of our picture of matter as representations and space as a state space arising as an epiphenomenon due to the differences in the representations, we can interpret the expansion of the universe as an evolutionary process of continuously increasing degree of dissimilarity among the totality of all representations. Indeed, if a set of representations become more and more dissimilar, the distances among them in state space would grow bigger and effectively the space separating them will expand.

In this way, we can interpret the observed expansion of the physical universe as a process of differentiation of the representations making up the particles of matter, which is very much akin to the process of accumulation of representations that we posited in relation to the theory of the evolution of life on our planet. The accumulation of representations in living organisms is a process of creation of novel structures, incorporating more and more of the inanimate particle structures into increasingly complex living structures. If we picture this process unfolding in the abstract state space of the states formed by the particle structures belonging to living organisms that behave like single units (biochemical molecules, molecular complexes performing a specific function, the particle structures in the brain engaging in a shared entangled state, etc.), then what we imagine would look very much like a miniature universe where new particles emerge everywhere resulting in an expanding state space.

We can also run this process backwards in our imagination and see how the state space shrinks all the way down to a tiny speck in state space which is basically the state of the first organic molecules formed in the primordial oceans of the young Earth. So, the expansion of the state space is actually a process of increasing complexity of the structures forming this state space, and it is indeed possible for such a space to have an origin represented by the simplest structure that has a state of the type that is characteristic for the universe formed by the states of all representational structures.

In that way, we need to regard the evolution of the universe as a process of increasing complexity of the material structures inside it, which is also equivalent to an increasing size of the universe. The state of lowest complexity, or the simplest configuration of matter, is a limit which cannot be crossed and represents the beginning of time and space, i.e., the origin of the universe. On the other hand, it is logically possible that there is no limit to how complex the representational structures can grow, since once they reach any given configuration, no matter how complex it is, it is always possible to transform to a novel configuration according to some rules of transformation, and this new configuration will be a successor to the previous one and therefore more complex in terms of the transformational rules. The only necessary condition is that the universe consists of a multiplicity of representations and not a single representation, i.e., its state does not revert back to a singularity.

As we said at the end of the previous chapter, the abstract state space formed by the states of living structures, and especially the conscious states of the human brains, looks like a miniature universe populated by a not very high number of unique particles corresponding to all possible mental states realizable in the currently existing human brains. Compared to the number of particles in the physical universe their number is indeed vanishingly small and the space that they form is very high-dimensional but with very limited extension, due to the fact that all particles in it are different and therefore cluster near the origin of the coordinate system of the space.

In that way, the abstract universe formed by the mental states looks like the very early stage of the existence of the physical universe, at a time point even earlier than the inflationary epoch which started about 10^{-36} seconds after the origin of time. So, we can regard the current organization of material structures into conscious living beings on the surface of the Earth as a very early stage of the origin of a new universe with possibly new rules for the interaction of matter inside it, since the particles of matter there correspond to conscious states, i.e., shared entangled states of physical particles, in our material universe and they do not need to obey the rules of physics that the material particles obey.

In order for the universe of abstract conscious states to expand and become as material as our three-dimensional universe, the physical embodiment of the conscious mental states that realize it would need to form some kind of substrate with highly regular structure which does not get affected by external perturbations that can disrupt the regularity of the structures. Basically, it means that all matter in the three-dimensional, physical universe should be incorporated in living structures sustaining conscious states and there should be no external material structures that can disrupt the functioning of the conscious ones. Stated crudely, it means that all matter in the universe should be incorporated in physical structures like the brain and there will be no other matter left, e.g. in planets, stars, black holes, etc.

This may seem like a far-fetched requirement, but if we consider the current evolution of material structures in the universe we can see that this is indeed the direction of the evolutionary process. Living organisms become more and more complex on the surface of the Earth and the emergence of consciousness results in a higher and higher impact on its environment—humans are transforming the surface of the Earth more and more to suit their needs.

If we extrapolate this principle to its ultimate end, we must conclude that at some point in the distant future all matter on the surface of the Earth will be incorporated into conscious structures. But this cannot be the end of the story since this does not look like an end state of time and the existence of the universe, so we have to presume that the process will continue, meaning that eventually all matter that makes up the Earth will be incorporated somehow into conscious structures. This cannot be the end of the story either, since we already are able to travel to nearby cosmic objects and these abilities would definitely be superseded by the future conscious life forms. Thus, we can project that not only our planet, but also the neighboring cosmic bodies will be transformed into conscious particle structures, and then the larger formations, such as the solar system, and the nearby stellar systems, and so on. Eventually, all of the matter in the observable universe can be potentially converted to conscious particle structures of uniform composition and this, then, is the logical limit of the evolutionary process.

In that limit the distribution of the particles that are familiar to us—mostly protons, neutrons and electrons—will be uniform and they all will be engaged in entangled states. Those states will be of the kind of the so-called quasiparticles that we encountered in chapter 1, and those particles would form the next level of reality and a new universe with different phenomenology and different laws of physics from the one we know.

The flow of time in this picture of the universe parallels the story of the evolution of space. As we saw, space has an origin in the Big Bang, expands continuously, and eventually a new origin emerges somewhere in the universe, and matter, and therefore space, starts transforming from structures of the old universe to structures belonging to the new universe. The same happens with time, since according to our discussion in Chapter 1 time is an epiphenomenon determined by the representational relations among particles and more specifically the difference components among them and the rules of their transformations. As the representational structures form higher-order entangled states, the rules governing the effective interactions among those higher-order states transform as well and eventually settle to a fixed set of rules which play the role of the laws of physics in the new universe. This settling process affects and determines the flow of time.

At the current stage of the evolution of conscious states the rules governing the thought processes realized by human brains are not very regular. These are basically the rules that cognitive psychologists, linguists and other cognitive scientists are trying to codify. This phenomenology, as we said, corresponds to a very early stage of the evolutionary process in the universe of conscious states, one which corresponds to an epoch earlier than 10^{-36} seconds after the origin of time in this universe. So, the flow of time in this new universe is extremely slow compared to the flow of time in the physical universe which engenders it.

The reason for this is that the material structures in the new universe are still very diverse and very few in number. As the universe matures, the structures become more regular and correspondingly the flow of time becomes more regular. In the limit when the material structures become very uniform and a very large number, the flow of time becomes as we know it—smoothly progressing at a constant rate in one direction. This is currently not the case with the evolutionary process of human thought, although there is already a discernable direction of evolution towards more complex and more regular states that represent better the external world.

We said that the rules that govern the interactions of the quasiparticles in the new universe will be different from those in the underlying physical universe, and it would be interesting to ponder in what way they will differ. They would need to be more complex, since the new representational states of matter will be more complex, so it is reasonable to assume that the state space they form will differ from the state space in the underlying physical universe, and the main way one space can differ from another is in the number of its dimensions. So, we can suppose that the new universe will have a different dimensionality, and since it ought to have more complex laws, this means a higher number of dimensions.

The Dimensionality of the Universe

In our discussion of matter in Chapter 1 we regarded space and time as epiphenomena arising from the ways the representations that make up all matter in the universe interact with each other. In that view, the dimensionality of space would also be a consequence of the rules of interaction and the specific internal structure of the representations. In some abstract sense the number of spatial dimensions reflects the number of ways a representation can transform while remaining a representation of another entity, i.e., something like the notion of degrees of freedom in physics. In our three-dimensional universe the number of dimensions, or degrees of freedom in the interaction processes among representations, arises from the particle property called spin, which was also discussed in more detail in Chapter 1. Particles with half-integer spin cannot occupy the same location in space and this is what gives rise to cosmic bodies and distances between them. In other words, if there were no particles with half-integer spin, the universe would have the extension of a single particle, since all particles that make up matter would occupy one and the same location.

So, we can suppose that the quasiparticles arising from the regular representational matrix arranged by the action of the conscious organization of matter in our three-dimensional universe will posses a property analogous to spin, but with more complex rules of interaction, and this would engender a four-dimensional epiphenomenal space in which those particles will be situated and will be interacting.

Thus, we can view the universe as a collection of a vast number of representations which posses a certain degree of complexity of their internal structure and interactions at any given moment of their existence. Their representational nature, i.e., the fact that they all reflect each other and in this way share the principles of their organization, results in conventional internal structure and rules of interaction of all representations making up the universe, effectively producing a highly regular epiphenomenal space and smoothly flowing time in which they are situated, as well as immutable laws of physics governing their behavior.

Eventually, in a place in the universe with high concentration of representations (stemming from a high diversity of representational structures and interactions among them) the degree of complexity increases beyond a threshold and results in a new organization of the representational structures—what we call the conscious organization of matter. This is the origin of new types of representational structures with new rules of interaction among them, i.e., a phase transition to a new, higher-dimensional universe. Effectively, this is a Big Bang-type of event which starts converting representational structures from the old order to the new, more complex order.

So, we can regard the Big Bang as a phase transition in the evolution of the universe where the number of spatial dimensions increases by one. We can suppose that the simplest state of the universe, which we can regard as the primordial Big Bang, engendered first a one-dimensional universe which became complex enough to engender a two-dimensional universe and the two-dimensional universe in turn engendered our three-dimensional universe after some time.

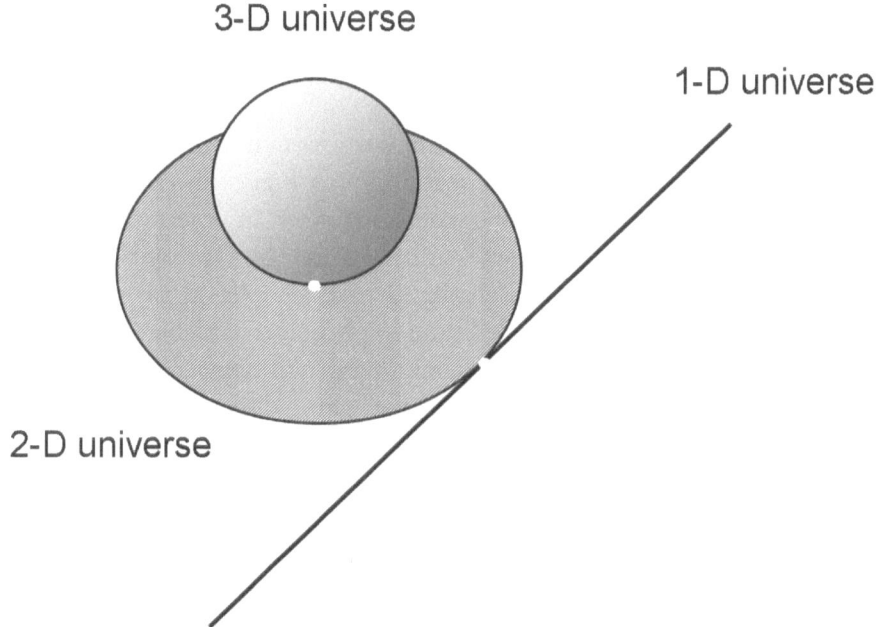

Figure 10: A schematic depiction of the three successive stages of evolution of the universe, each with a higher dimensionality than the preceding one. The white dots represent the Big Bang events.

It should be noted that the creation of a new universe in a Big Bang occurs while the old order is still in full swing, which means that most of the existing representations still behave according to the rules of the lower-dimensional universe. In fact, the transformation process does not need ever to be completed and there may be remnants of any of the earlier universes in the subsequent universes. So, this means that some matter in our three-dimensional universe may behave according to the rules of the two-dimensional or even the one-dimensional universe preceding it.

If that were the case, what would be the effect of this primordial matter on the particles in our three-dimensional universe? To answer this question we can try to imagine the effect of inanimate matter on the conscious states in the human brain, which we regard as particles in a higher-dimensional universe. In this case we also have particles organized and behaving according to the rules of a lower-dimensional universe affecting the states of particles in an epiphenomenal higher-dimensional universe.

To begin with, we can observe that the particles in the higher-dimensional universe, i.e., the conscious states of the human brains, interact only with entities of the same kin—with other conscious states. In other words, we can think of and imagine other people's mental states, but we cannot imagine any possible non-human shared entangled particle states which would play the role of particles of another type in the epiphenomenal higher-dimensional universe. Thus, particles in a universe of a particular dimensionality interact directly only with other particles of the same dimensionality and are influenced only indirectly by particles with a lower dimensionality.

The indirect effects can be illustrated as well by the way the inanimate matter shapes our conscious states. Basically, the conscious states are grounded in the perceptual interface with the external physical environment, which is part of the three-dimensional universe, and the relationships among them reflect the phenomenology of the three-dimensional world. Most of our knowledge is made up of concepts reflecting the behavior of macroscopic objects in the three-dimensional universe or their group behavior at a more abstract level. So, the particular configuration of the particles in the epiphenomenal higher-dimensional universe and the rules of their interactions are determined by the material structures in the lower-dimensional universe in which they are immersed, but the internal structure of the particles is not, just like the content of a human conscious state is determined by the peculiarities of the entanglement of particle structures in the brain and not by the particle structures in the external, three-dimensional world.

In summary, it turns out that the effect of lower-dimensional particles on the higher-dimensional particle structures in the epiphenomenal universe is only indirect, and can be felt as influences on the configuration and movements of the particles but not on the representational contents of the particles and their interactions. That is to say, according to our picture of a particle as a representational complex, that the representations determining the internal structure of the particles and the rules of their interactions, i.e., the laws of physics, do not exchange information across dimensionalities, i.e., three-dimensional representations interact directly only with three-dimensional representations but not with lower-dimensional representational structures.

In the language of physics this means that the lower-dimensional material structures cannot be engaged directly in interactions with three-dimensional particles and therefore they are not particles in our three-dimensional universe, but only affect the behavior of the three-dimensional particles. This may seem a rather odd proposition, but it fits exactly the description of the mysterious dark matter introduced in cosmological theories in order to explain some discrepancies between the observed and calculated behavior of matter on the large cosmic scales. Dark matter does not interact (or interacts extremely weakly) with ordinary matter (that is why it is "dark"), but its presence can be felt through its gravitational effect, i.e., it interacts with ordinary matter only via the force of gravity.

It seems fitting that the lower-dimensional particles hypothesized within the framework of our picture of matter as representations are a candidate explanation for dark matter in modern cosmological theories because they posses also some other features pertinent to dark matter. The lower-dimensional particles in some sense are in close proximity to the higher-dimensional particles because of the grounding of the conscious states in the perceptual states in our analogy. This corresponds to the observation that dark matter clusters in the regions with ordinary matter (and vice versa), near galaxies and galaxy clusters. Also, in terms of its gravitational effect dark matter is estimated to be much more dominant in the universe, accounting for about 25% of its material content compared to less than 4% belonging to ordinary matter. This also fits our picture of a higher-dimensional universe arising within a lower-dimensional universe and gives an idea of the relative sizes of the two. It also means that dark matter gets gradually transformed into ordinary matter and their ratio changes with time.

Another line of evidence in support of our hypothesis that particles in the three-dimensional universe originate from lower-dimensional material structures comes from the picture of matter provided by string theory. In this framework, particles are regarded as vibrating strings, which are one-dimensional entities that reside within a single or between two different two-dimensional sheets called branes. This theoretical framework succeeds in describing large part of the phenomenology of particle physics in a parsimonious way, but it fails in some regards and therefore is not regarded as a proper, verified theory that could serve as the fundament of all other theoretical frameworks in physics.

Nevertheless, viewing particles as one-dimensional entities at extremely high energies and extremely small spatial scales fits our idea of matter forming universes of successively higher dimensionality. The energies and spatial scales at which the string nature of particles can be observed belong to a time interval which is extremely close to the origin of time at the Big Bang—somewhere in the vicinity of 10^{-43} to 10^{-36} seconds after the Big Bang. This is a time frame prior to the inflationary epoch, which in our picture of the evolution of the universe corresponds to the lower-dimensional phases of organization of matter. So, it seems fitting that the mathematical models describe particles as one-dimensional entities at very high energies and small spatial scales—this is a state of organization of matter which belongs to the one-dimensional or the two-dimensional epochs preceding our three-dimensional universe, according to our view of matter as representations.

There is one final note we need to make regarding the terminology we use when talking about the Big Bang. In standard cosmology there is only one Big Bang, which is the origin of time and of our three-dimensional universe, but in our theoretical framework of matter as representations there are successive stages of existence of universes with successively higher dimensionality and successive events of origination of universes of a higher dimensionality, each of whom we may call a Big Bang. So, there is one primordial Big Bang, which is the origin of time and gives rise to a one-dimensional universe populated by primordial strings (which, however, interact in simpler ways than the strings in string theory, otherwise they would behave like three-dimensional particles on the larger scales and would form a three-dimensional universe), and after that there are two more Big Bangs which produce a two—and a three-dimensional universe respectively. In terms of the standard cosmological picture, the third Big Bang corresponds to the inflation event occurring between 10^{-36} and 10^{-32} seconds after the primordial Big Bang in the standard cosmological scenario[9] and the second Big Bang is somewhere in between

---

[9] The reason why the inflation event in the standard cosmological picture corresponds to the birth of the three-dimensional universe in our representational picture of the world is that the description of the inflation event looks very much like the description of the phase transition from a lower-dimensional universe to a higher-dimensional one which we conceived in some more detail earlier in the section. In the standard cosmological picture inflation is a result of the negative energy engendered by the uniformity of the quantum field that filled up all of space in the pre-inflationary phase. In our picture the expansion of space is due to the uniformity of the representations, i.e., to the dominance of the similarity component over the difference component. As we said earlier, the two limits – of total similarity and total difference – are unreachable, since complete similarity between two representations means that they occupy the same place and behave in exactly the same way, therefore they can form effectively a single representation, not two different ones, and complete difference means that the two representations have no chance of ever interacting, meaning that they belong to two disjoint universes, and there is no reason to assume the existence of other universes which have absolutely no effect on ours. So, the components of similarity and difference can get arbitrary small, but cannot vanish in a single universe made up of representations which have at least some shared history of interaction, which would be necessary the case if we assume that the universe originated from a single state in a primordial Big Bang event. The process of accumulation of representations leads to constant increase in the similarity component on average (although there are deviations from the average in both directions), leading eventually to highly similar complex representations which are able to sustain higher-order representations (i.e., quasiparticles in physical terms) that constitute effectively a new universe with higher dimensionality. The reason why highly similar complex representations engender extended dimensions of space (akin to the inflation of space due to the negative pressure of the uniform quantum field) is that very similar representations can be ordered in a way that ensures gradual transformation of their structure (just like slightly different numbers can be sorted in ascending order and thus would populate a number line signifying a single dimension in geometric space). Dissimilar representations, on the other hand, would tend to vary independently of each other, which would mathematically be represented by independently

the other two.

Why is the Universe So Big

As we stated earlier, the vastness of the universe is one of its most striking features to us, human beings on planet Earth. We also said that the human scale is about twice smaller than the scale of the universe as it is bigger than the scale of the subatomic particles. While we have developed tools with which we can manipulate individual particles, we can only passively observe the most distant objects in the universe and try to discern some of their properties based on the spectral composition of the light they emit. So, the vastness of the universe remains more difficult to grasp and rationalize than the oddness of matter at the very small scales.

---

varying numbers, and in mathematical terms the independent components of a vector represent different dimensions of geometric space). So, the picture of the universe made up of highly dissimilar representations looks like the picture of mental space made up of all concepts (i.e., well-defined states of the conscious state realized in the human brain) and they correspond to a physical picture of dissimilar particles occupying nearly the same location in a very high-dimensional space. This single location corresponds to the singularity state of the universe in the pre-inflationary epoch. Eventually, when the conscious structures engulf all physical matter in some region of the universe and become more and more similar, there is a phase transition event when they become maximally similar and form extended dimensions of space due to the fact that they populate evenly the state space of all possible configurations they can take and there is a way of ordering them in smoothly transforming variations of their properties. Still, there are small differences in the smoothness of those variations, due to the vestiges of the degree of dissimilarity among the representations, and these correspond to the quantum variations in the inflationary field that result in the uniqueness of the material structures in each region of space in the universe. In this way our representational picture of the transition from the old type of organization of matter forming a lower-dimensional universe to a new type forming a higher-dimensional universe corresponds closely to the picture of the inflationary event in standard cosmology.

The most straightforward answer to the question of why the universe is so big is that it has been continuously expanding since the Big Bang and has done this for a very long time—about 13.7 billion years on the human scale. On the scale of subatomic particles that makes about 10^{32} oscillations of the particles' states, if we assume that the dominant frequency of oscillation of the particles is that of visible light. If we also assume that particles exchange photons that frequently (which is simply a guess), then given that the number of atoms in the universe is estimated to be on the order of 10^{80}, the number of interactions among particles since the origin of the universe (or at least since the creation of particles) is greater than 10^{110}! That is a rather large number, showing that the present distribution of matter in the universe and, consequently, its size are a result of a very large number of interactions among the particles of matter.

In our picture of matter as representations and space as a type of state space an expanding universe means also that the material content in terms of particle structures should be growing more diverse, i.e., in order for state space to expand its content needs to differentiate continuously its states. Given that state changes occur through the exchange of particles that carry the four fundamental forces, as we saw in chapter 1, this means that a continuous current of interaction events among the particles making up the universe would result in a continuous transformation of the states of those particles. Due to the nature of state space, namely that each point of this space represents a different state, a continuous transformation of the states of a set of interacting entities would most likely result in a diversification of those states and consequently an expansion of the state space, since any interaction event that makes two particles more different, even if it occurs by chance, would expand the state space, while any event that makes them more similar would probably not result in shrinkage of state space since there are likely to be other particles that occupy the more different states and in this way keep state space expanded. Basically, any state that is reached by any particle leaves a trace in its subsequent states and those of all particles with which it interacts, so that this state is reachable more easily in the future and therefore may be considered as a part of the state space which cannot be lost. This mechanism allows state space only to grow, but not shrink. It is similar to the picture of a gas expanding in an empty volume.

This picture of continuous diversification of states and expansion of state space in fact reflects the familiar law of accumulation of representations that we stipulated as the driving force of the evolutionary process culminating in the emergence of the conscious organization of matter. Indeed, representations accumulate not only in the states of individual particles, but mainly in the collective states of particle structures, and this is where the greatest diversity of states continuously arises. The creation of diversity, thus, would be greatest in particle structures that have a high degree of interactions internal to the structure and a low degree of interactions with external structures. Well, this is precisely the case with most material structures in the universe, which, as we well know, have collapsed to form cosmic bodies, such as stars, planets, black holes, etc., separated by large expanses of space between them.

So, we can conclude that the lumpy distribution of ordinary matter in the universe in the form of cosmic bodies is what drives the expansion of space. Just from the fact that space is a kind of state space we reach the conclusion that it must be expanding and its rate of expansion would get higher the lumpier it is. This fits the current observational evidence of an accelerated expansion of our universe.

There is another conclusion we can make from our picture of the expanding state space regarding the properties of its expansion. If particle structures in the form of cosmic bodies develop ever more different collective states, it means that the space between them would be expanding, but the space that they occupy and its vicinity would be expanding at a lower rate, if at all. So, the expansion of space is highest in the emptiest regions of space, those farthest away from any material structures, and lowest in the space occupied by material structures. This matches precisely the description of the action of the mysterious 'dark energy' in standard cosmological theories, which is supposed to arise from the vacuum and have a repulsive effect on all material structures.

So, left on its own devices the universe would expand continuously driven by the continuous accumulation of new representations in terms of complex particle structures. This picture, however, does not account for one very important phenomenon that is part of the universe, namely, consciousness. As we saw already, the conscious organization of matter is a special one and its tendency is to grow and engulf more and more of the material structures in the universe. In its limit, this process will eventually engulf all matter in the universe and give rise to more complex representational relationships, which will effectively constitute a new, higher-dimensional universe with new kinds of particles and laws of physics.

Since conscious material structures are bound in a shared entangled state they do not have the tendency to differentiate their states, but rather the accumulation of knowledge in terms of complex representations spanning the entire conscious particle structures would have the effect of increasing the degree of similarity over the degree of difference of the representational structures and thus effectively contracting the state space of the material structures. (Note, however, that the state space of the conscious mental states may at the same time increase, leading to expansion of the higher-order state space and its respective universe.)

So, we can conclude that the action of consciousness is exactly opposite to that of 'dark energy', or the natural tendency of material structures to differentiate their states in the process of accumulation of representations in particle structures that are not involved in sustaining conscious organization of matter. In other words, we can consider 'dark energy' as having a repulsive effect on inanimate particle structures and the opposite—attractive—effect on conscious particle structures. Given that at the present moment the conscious structures are a negligible fraction of all particle structure, the attractive effect is correspondingly minute and the repulsive effect is dominant, causing the universe to expand, but with the growing influence of the conscious organization of matter this trend will gradually reverse (in the course of billions of years) and eventually the three-dimensional universe will stop growing and start shrinking, or rather, transforming into a four-dimensional universe. As matter transforms from lower dimensional structures to higher-dimensional ones, the lower-dimensional universe will gradually shrink, while the higher-dimensional one will expand.

The vastness of the universe, or rather, the expanses of (mostly) empty space between the concentrations of particles making up the cosmic bodies, as we saw, is a result of the action of gravity, or the ability of particles to retain some 'memory' of their earlier states which produces a tendency for their states to return to earlier configurations, thus forcing them to clump in state space. Over the long time of the existence of our three-dimensional universe this tendency has made the distribution of matter in the universe at the finer scales very uneven—with particles packed very tightly in some small regions of space and vast expanses of almost empty space in between.

This arrangement of matter has one fortunate consequence. Any particle structures that may form with a growing degree of representational relations among them, which we deemed as the hallmark of life and the conscious organization of matter, would be forced to interact locally, in the vicinity of the clump of particles making up the cosmic body where they formed, for a prolonged period of time before they spread to other cosmic bodies or systems. This ensures the high degree of reconciliation of the conscious particle structures leaving one local region of the universe to spread to other neighboring regions and in this way increases the sophistication, and therefore the chances of success, of the diaspora of conscious structures.

In other words, the lumpy distribution of matter in the universe is beneficial for the successful realization of well-behaved higher-order particle structures (quasiparticles) by allowing them to form in one place and get reconciled before engulfing other lower-dimensional particle structures. If the distribution of matter in the universe was less lumpy, the higher-order structures might spread before they get well reconciled and this could lead to conflict on the large spatial scale, i.e., to violent negative experiences for the most advanced structures of consciously organized matter. In a sense, the vastness of the universe ensures the relatively peaceful transition to higher-dimensional structures associated with mostly positive conscious experience for the conscious particle structures.

Our discussion of the evolutionary process of the universe thus far was focused mainly on the past and the present. The understanding of the evolution of the universe from its hypothetical beginning to the present moment is based on empirical evidence—we can observe the universe as it is now and we can also peer back in time by observing very distant objects, since it takes a lot of time for light to travel the distances from them to us. For these reasons scientific discussions are limited to events in the past that can be inferred from observations in the present, such as the microwave background radiation, but there is very little discussion on the future of the universe, since it cannot be readily inferred from observations. The main theoretical propositions regarding the future concern the very distant future, i.e., the end state of the universe, and are based on observation of the strength of the gravitational force acting on all matter in the universe, leading to three distinct scenarios—a collapse of space back to a singularity, a steady state, and an eternal expansion resulting in a cold and lifeless universe. More recently, a number of cosmological theories have been proposed advocating the possibility of a cyclical nature of the evolution of the universe—going through successive phases of expansion and creation events in recurring Big Bangs. This is pretty much all that modern science can say about the future of the universe.

Our understanding of the representational nature of matter and the ensuing role of the conscious organization of matter in the evolutionary process of the universe allows for the development of a much more detailed picture of the future developments in the course of evolution of material structures in the universe. Such an understanding would be based on imagining all matter, including living and conscious structures, as representational complexes. In that framework there is a clear and smooth progression in the evolution of the representational complexes informed by our hypothesized law of accumulation of representations. When extrapolated into the future, this law and the corresponding understanding of its manifestations in the organization of material structures can enable us to picture the major milestones in the future course of evolution of the universe based on observations of the present, much as from the observations of the present we infer the occurrence of events in the past. Logically, there is not much difference in those two types of inference. The main requirement for being able to infer facts about the future is to have completeness of the knowledge of the present at some level of detail. If our knowledge of the present is complete within some conceptual framework, we can run the model forward and observe future developments with a certainty corresponding to the confidence that we have that our conceptual framework is an adequate representation of reality.

The following sections will deal with such a conception of the future based on our insight that the evolutionary process of conscious material structures leads to a new Big Bang engendering a higher-dimensional universe. This is the horizon of what we can possibly know about the future given our present knowledge. The other limitation is that we can imagine only those future developments that can be expressed within our current conceptual apparatus, i.e., we can imagine only what we would be able to understand given our current level of knowledge. Undoubtedly, there will be many developments in the future, especially in the more distant future, which would be beyond our comprehension, but at the least we should be able to conceptualize the most general processes as variations of the process of accumulation of representations.

The Future

500 years into the future

During this time period of approximately 500 years ahead from the present there will be significant developments in terms of science, technology, social order, and other changes in the major components of modern life, but life on Earth will retain its basic form and structure from today and the past few thousands of years of human civilization. The advances in science and technology would probably lead to the emergence of simple artificial intelligence in the form of robots capable of performing simple human tasks of the kind employed nowadays in manufacturing and agriculture. Robots will be able to grow food, process materials and produce final goods, and people will be focused on tasks requiring higher intelligence, like scientific discoveries and building even more complex robots capable of doing even more complex work. The increase in the capabilities of artificial machines would be directed towards conquering more and more different habitats than the one suited for humans—the dry surface of the Earth. The main effort would be to build robots capable of operating in very different environments like the oceans and the surfaces of other planets in the Solar system. Those robots will be made in part with nanomaterials which would bring them closer to the structure and functioning of an organism, although they would still not have reached this stage. On the other hand, human bodies will be more and more malleable and integrated with technology. It will be possible to prevent and cure most well-defined diseases and there will be sophisticated methods of influencing the functioning of the brain, leading to more successful education and higher levels of intelligence. The major energy source will be the sunlight and it will be used to transform materials into finished products in many ways. There could be also thermonuclear energy available, which would help make transport and travel into space commonplace.

There will be major social changes as well. The focus of social life will shift from making money and devising schemes for redistribution of money to acquiring knowledge and building more sophisticated devices which will be capable of transforming the environment more successfully according to the needs of the people. Life will become much more organized and people will plan not just the major direction of their personal life and career but also details about what exactly they will do and even what they will think far ahead in the future. It would be commonplace to think of one's entire life as a whole in terms of what one is going to accomplish and create, and to manage it on that scale. People will see themselves less as individuals, and more as propagators of a larger process of transformation of the material world around them. Life expectancy would be somewhat higher than today, but not radically different. The quality of life, though, would be much higher due to the more successful elimination of negative experiences and the prevalence of positive experiences based on the higher level of intelligence. We could say that in terms of representations, the structures in the human brain producing the conscious state would be in more harmony with the environment (including other brains) which would lead to more frequent matching of the representations producing positive feelings and less frequent conflicts of representation producing negative feelings. Basically, life will bear a general similarity with today, with people living in buildings, working, and engaging in enjoyable activities during the free time, but it all will be operating much smoother, with more intent and more understanding, and with a better awareness by everyone about the whole process. This would mean more agreement and less conflict, which would be restrained to decisions affecting other individuals and will be resolved easily.

One thing that people will strive for but might still not be able to achieve is to build artificial consciousness of the type existing in the human brain as described in Chapter 2. Hopefully, there will be a consensus regarding the nature of consciousness and its basic principles of functioning (and they hopefully will be the ones described in this book!), although there will still be no direct proof of this in the form of the ability to create or manipulate an artificial object endowed with consciousness. However, the science at that stage would be making fast advances and with the help of the artificial intelligence devices based on a mix of quantum and classical computation it would be possible to prepare material structures which come closer and closer to being able to maintain a conscious state. Importantly, they will not only replicate the structures in the human brain with carbon-based biochemistry, but also other possibilities will be discovered for creating complex adapted structures based on different biochemical materials. The difficulty in replicating consciousness is the vast number of possibilities for arranging different interacting chemical structures and only some very special arrangements, namely those that are highly adapted to each other, would turn out to support consciousness. This feat, however, would hopefully be achieved within the next few hundreds of years and it will be reality in the next time period that we will discuss—1000 years into the future.

1000 years into the future

The civilization around year 3000 will be still based on robots and the struggle will be to endow them with consciousness of the kind that humans possess. This feat can be achieved by a gradual process of discoveries of adapted structures which can be put together at the molecular level using techniques from a more advanced nanotechnology. This process will resemble the natural evolutionary process, but it will unfold at a much faster rate due to the selective action of human consciousness. It will lead not only to the replication and manipulation of human biochemistry, but also to discovery of other possible biochemistries enabling consciousness based on a different mix of chemical compounds.

This breakthrough would open entirely new possibilities for the future humans. Now it will be possible to build and send much more intelligent robots to distant planets, which would be adapted to the physical environment there and could essentially colonize the new habitats and start to transform them. The goal of this transformation will not be to make it suitable for humans to live there, but to extract raw materials and build finished products which are needed by humans. On the other hand, humans on planet Earth will become even more integrated with the inanimate intelligence of robots through all kinds of human-machine interfaces and their bodies will be even more transformed and manipulated in order to achieve this integration. One consequence of this advanced technology would be the virtual elimination of illness. Human bodies will be constantly monitored and regulated so that they do not develop severe imbalances or dysfunctions manifesting themselves as disease. The only health problem will be trauma, but it too will be highly treatable except for the most severe cases. It may be even possible to replace the whole body, part by part, with a younger one, but it will not be possible to rejuvenate the brain, so ultimately people will die probably in the early decades after 100 years of age. Some people may long for immortality, but others may consider this an illusion, since they would realize that any rejuvenation of the brain would entail a modified individuality which is essentially undistinguishable from the death of one individual and the birth of a new one. So, when immortality eventually becomes possible in the following centuries it will not be the kind of amazing feat that it seems nowadays but it will be considered not much different from the cycle of birth and death.

The people 1000 years from now will be very much concerned with the future. They would realize that it is possible to merge the natural human consciousness with the artificial consciousness of robots realized in a different biochemistry and they would be hard at work on achieving that. This would enable the transfer of individuality from one body to another, even across different biochemical structures. However, the transfer of individuality is a very complex process, which means that a huge number of structures at the atomic level need to be replicated from one brain to another and initially only a small part of them could be transferred, leading to replications of only basic consciousness capabilities. This would create a two-tier society of creatures with very low intelligence, but who can transfer their individuality from one body to another and thus live eternally, and creatures with high intelligence which would follow the normal cycle of birth and death. This, however, will be a feature of life in the next time period—5000 years into the future.

5000 years into the future

The jump from 1000 to 5000 years ahead is much bigger than the one from 500 to 1000 years, so the changes during this time period will be more profound and encompassing, leading to a whole new outlook of life in the year 7000. Human society at that time would have barely anything in common with the way we know it now. Humans will be much more like the cells in the brain—locked in place and heavily connected to their environment, exchanging information with it and performing almost exclusively decision making and intellectual work. They will stay mostly at home, although this will be a very different type of home, one where their brains are connected directly to actuators in the outside world that do the physical transformations of materials. Leaving their seats would be both cumbersome and unnecessary, because their bodies will be kept in excellent condition by the technology which makes up most of the home and their minds will be in direct causal contact with objects in the environment, enabling them to see and feel distant locations as if they are there. The surface of the Earth would look like the body of an organism, with lots of structures for capturing energy used to transform different materials and lots of pipes for distributing these materials, but few freely moving objects. Only robots for maintaining the infrastructure and doing new construction will be wandering around. However, lots of buzz and activity will be going on in a different place—in the newly established colonies on the other planets and their satellites in the Solar System. This activity will be also the main preoccupation of the humans populating the Earth. A relatively small proportion of them would be needed for the maintenance and transformation of the surface of the Earth and a relatively larger proportion would be engaged in transforming the extraterrestrial habitats. The transformations there will be performed mainly by robots possessing some limited degree of consciousness and intelligence, and the task of the humans will be to supervise and direct their work. The robots will be working mainly on making the environments of the other planets suitable for life for some of the types of artificial conscious beings created by humans, not for humans themselves, although there may be some small number of colonies which will be adapted also for humans. In summary, the conquest of the Solar System will be the main preoccupation at that time period, but also some exploratory missions will be launched to the nearest stars outside the Solar System.

The personal life of a human will also be very different from what we are used to nowadays. It will be much more orderly and predictable, following a well-controlled path from childhood to seniority, with a consequence that time would seem to flow faster and the life experience would seem shorter. To this phenomenon will add the fact that human consciousness will be somewhat expanded due to all the integration with peripheral devices and the constant contact with other consciousnesses. So, naturally humans will long for a longer life and as mentioned earlier they will try to transfer their individuality into another conscious system. Basically, the close contact with other brains through the means of technology will allow for partial merging of individuality of conscious systems with different age and even with different biochemistry. This would lead to a smearing of the borders between different individualities and the perceived gradual transfer of the individuality (albeit incomplete and imperfect, not an exact copy) from one conscious system to another. For some, as we said, this may be highly valued achievement but for others it would not be that important. The high diversity of conscious beings and environments in which they operate would create more diversity in ideas and thought processes, counteracting the natural phenomenon of increased cohesion due to the high degree of interconnectedness of conscious systems. These two opposing tendencies may lead to something like speciation of conscious beings, leading even to conflicts between the different species of creatures. However, these conflicts should incur minimal violence and should be short-lived. Just like in today's society, there will be different degrees of sophistication and development of the individuals which form the basis of the conflicts. It seems logical, though, that all conscious beings would realize what they are and what their destiny is, so their relationships will be dominated by cooperation most of the time.

10 000 years into the future

As we said, time would seem to pass faster at this time period meaning that changes would take more time and would be more gradual and incremental, so major developments encompassing the whole system of living beings would be rarer. Around year 12 000 the processes started 5000 years ago will be in full swing and life would have expanded more quantitatively than qualitatively. The surfaces of most planets in the Solar System will be colonized to a good degree and they will have their own ecosystems and idiosyncratic populations. By that time the Sun would have become somewhat hotter, so the climate of the planets would have changed a little bit, also on Earth. Earth's surface, though, will be well-engineered to accommodate to this change and to absorb the increased amount of energy coming from the Sun, so this phenomenon will be welcomed rather than endangering the human population. Humans at that time point would be completely unrecognizable, virtually like cogs in a big machine, living in bubbles of advanced nanotechnology without moving out of them, much like the cells in an organism. It seems even questionable whether they will have bodies at all or just the brains will be left immersed in some nourishing and communicating shell. There will be very little left of the individuality that we have today and people's minds will organize themselves into some kind of swarm intelligence as it exists in social organisms like ants and bees, but at a much higher degree of consciousness. This would not seem such a radical change if we consider how much human consciousness has developed in the last 10000 years, from the time of the first tool-using humans whose consciousness must have been halfway between the animal and the human one. This expanded consciousness and loss of individuality would make time seem to go even slower for this super being, which would have the effect of making distances appear shorter and the Solar System will not be such a big place anymore. Also, the nearby planet systems will be in the initial stages of colonization and although it takes several years for signals to travel to them at the speed of light, the communication with them would not appear that slow or cumbersome. This time period will mark the beginning of the colonization of the Milky Way galaxy building upon the successful colonization of the Solar System which is in its apex.

Consciousness, as we said, will be much more distributed in a network of well interconnected brains but this does not mean that it will effectively form one giant brain. We saw that the conscious process arises at the interface of two representations formed physically by resonating atomic structures. This is possible in a brain because all atomic structures in it are adapted to each other and they are in constant physical contact. Even the very sophisticated communication systems 10 000 years into the future would not be able to achieve that degree of interconnectedness and density of representations, so the conscious state will remain confined to a single brain but it will be easily influenced by the conscious states of other brains much like in a fictional telepathic communication or in a very intense conversation which can be experienced nowadays. This means that the content of thoughts will be similar in its nature to the thought process that we have now, i.e., people will use some kind of more advanced language to communicate and the conscious state will have the pulsating form with interchanging coherence and decoherence states which we discussed in Chapter 2. This will also be the form of the conscious process in the artificial conscious beings based on a different biochemistry than the human one, although it may be possible that gradually transforming conscious processes will also exists, something like conscious plants, which will not have the pulsating character of human consciousness but will evolve their conscious state very slowly. This network of brains and conscious processes will prepare the way for the next development, which will be the hallmark of the next time period under consideration—50 000 years into the future—and namely, the integration of the network into a giant superorganism made up of the entire biosphere of the planet.

50 000 years into the future

We will make again a large leap ahead and imagine Earth after 40 000 more years of evolution. This is just about enough time for the development of a new species on the traditional evolutionary time scale and it is about the length of time with virtually no evolutionary change in humans, judging from the size of the brain. This time, however, the new species will be the giant superbeing endowed with consciousness, encompassing the entire biosphere of the planet (or maybe just the highly integrated part of it). It should be noted that as a percentage of the mass of the Earth the biosphere will be still an extremely minute fraction and it will not even cover the entire dry surface of the Earth, but it will be continuously growing and transforming the rocks from the dry surface and the water from the oceans into biomass. The conscious biomass—the brains—will be in turn a minute fraction of the total biomass, most of which will be concerned with capturing energy and transforming materials, much like in any animal body. Most of the activity of the superbeing will be focused on communicating and exchanging materials with extraterrestrial colonies. The colonies themselves will be on their way to transforming into superbeings, but would not have reached that stage yet. The main mechanism for transporting materials to and from space will not be space ships launched from the Earth's surface, but protrusions, or cables, extending from the Earth's surface to space stations in orbit, from which the payloads will be transported with rocket-propelled space ships. Initially the cables will be extended probably from the highest points on the Earth's surface—the mountain peaks, but later the technology may become sufficiently advanced in order to produce artificial peaks or stronger cables which can start from any point on the surface. This main mechanism for transporting material will become more and more significant in shaping the planet (and also all other planets and their satellites) and in determining its ultimate fate, as we will see later on.

 The biological organisms, as we saw, will be organized by consciousness into one superorganism with high level of integration of its biostructures (probably still based on cells and biomolecules like proteins and DNA as we know them today). However, this superorganism would not be able to achieve single unified consciousness, since this would require entanglement on a much larger scale than in the brain, so it will possess multiple, highly interacting conscious states, somewhat like particles in a fluid. They will be realized not only in the traditional biochemistry of the human brain but also in the other artificially created biochemistries and computational devices based on phenomena of entanglement in other material structures. Some of the conscious states, thus, will be realized in physical systems existing at very low temperatures, close to the absolute zero and the conditions in space. Since coherence among particles is much stronger at low temperatures, these conscious states will be developing much slower and will be simpler than the high-temperature ones, but their advantage will be that they are more regular and predictable. Thus, they will be suited for the slow processes of transforming the conditions in space on a large scale, which affect the gravitational interactions between the bodies in the Solar System and in this way can transform its fundamental structure and functioning. Overall, the changes in the Solar System as a whole will be that more and more material and physical structures will enter the interplanetary space and also more gases will be emitted in the interplanetary space from the propelling of the space ships. This tendency will smear a little bit the sharp division between large material bodies and empty space between them and will make conscious matter more uniformly distributed in space.

100 000 years into the future

At this point in time the changes relative to the preceding time period under consideration will be more quantitative than qualitative. The conscious superorganism on the surface of the Earth will reach even greater degree of integration and would be able to perform the transformations of materials at a faster rate. The transportation of materials to and from space will intensify and will reach a scale where the total mass of the planet may be significantly affected. At that time, the Sun will be hotter than it is now and the temperature on the Earth's surface will rise by a few degrees. The ability to change the mass of the Earth would allow the superorganism to change the orbit of the planet, moving it further away from the Sun if necessary. Thus, the climate of the planet can be kept suitable for the traditional consciousness-supporting biochemistry near 37°C. The superorganism will cover a significant proportion of the dry surface and it will use large quantities of water from the oceans to grow its biomass. The protrusions used for transporting materials to and from space will look more like mountains, with a solid base at the bottom and thinning out as they rise for about 100km towards space. This would make the planet look somewhat like the picture of a virus on the atomic scale, with the water gathered at the footing of the protrusions and the biomass covering them. The other bodies in the Solar System will start to look the same way, with the exception that many of them will not have liquids on the surface. The sending of materials into space will reduce the mass of all planets and satellites, while the interplanetary space will become regularly populated by colonies with conscious life forms. Many of these colonies will be launched to other planetary systems in the galaxy in order to colonize them. Since this would result in a large loss of mass for the Solar System, material will be transported back from the nearest planetary systems, which are on the order of 10 light years away. About that time most of the galaxy, which is about 100 000 light years across, will be reached by colonization missions and the transformation of the planetary systems in the galaxy akin to the pattern in the Solar System will be well under way.

The consciousness of the superorganism will be more advanced than that of a human, but it will still be operating on the same principles, only on a larger scale. The proper conscious matter, the counterpart of a human brain, will be larger than the average human brain nowadays, but it will probably not be much larger than a few cubic meters and weighing more than a few tons, since it still would need to consist of atomic structures adapted to each other, and their accumulation is a slow process. Still, the larger structure would mean slower evolution of the conscious state and therefore slower thought process. Considering the gap from the scale of atomic oscillations to the human scale, which is about the same as the difference from the current scale of human consciousness and the one we project for that time period (one atomic vibration, lasting on the order of a few hundred femtoseconds, is to one second what one second is to 100 000 years), we can see that it is not that farfetched to suppose that the conscious process will slow down so much as to make astronomical distances seem ordinary.

There will be numerous such superbrains dispersed regularly inside the biomass, and the rest will form the body of the superorganism whose main purpose will be to transform matter, starting from the chemical compounds of the rocks and water and producing end materials needed to sustain the life of the organism and to build different types of conscious organisms which can be sent to colonize distant planetary systems. There will be very little inanimate technology, as we know it nowadays, and most material objects will be either biological organisms (a huge variety of them, with different biochemistries) or inanimate matter in the form of raw materials making up the bulk of the planets. There will be still solid structures made of carbon or metal, but these will be part of the organisms, just like the skeleton is part of the animal bodies today. So, the natural world will look more like what it used to look for most of the history of the planet, with a sharp division between living and inanimate objects, in contrast with the modern world, where we have a lot of machines which are somewhere in between the living creatures and the inert inanimate compounds. From that perspective, today's technology turns out to be an intermediate stage of organization of matter, when intelligence (i.e., consciousness) in the form of highly organized matter shapes the unorganized matter, but still not to a degree which would make it organized enough to become conscious in its turn. With the advancement of intelligence the transformational capabilities of conscious matter will grow and eventually reach the stage where it can directly transform matter from the unorganized state to the organized state and incorporate it in the living organisms, very much like plants take in minerals and gases from their environment and convert them to biomass. Technology may still play a role in the colonization of other planetary systems if they are different from the known ones and there is still no suitable biochemistry invented which can survive the conditions there, but it should be only a matter of time before one is created.

500 000 years into the future

This amount of time would be enough for the full colonization of the galaxy and the transformation of the planets in the planetary systems from giant balls of particles to spike-dotted balls covered by a layer of biomass. They will grow smaller and smaller in size as material is constantly being ejected from them into interplanetary space. The exhaust gases from the transportation activity in space will form a halo around the cosmic structures in interplanetary space which will be gradually blown off into deep space by the solar wind. In the Solar System there may be technology to capture and reuse some of it in the form of a screen wrapped around the plane of the planetary orbits, enclosing all life in it. This screen could also collect energy from the Sun more efficiently and become a major energy source. The Solar System will start to look more like its initial phases of formation, with a large quantity of material between the planets, but this time instead of condensing to planets by the force of gravity the material will spread out under the influence of conscious intelligence. The Sun will be even hotter at that time and will provide more energy to the biomass, and maybe it would have become possible to create consciousness based on biochemistry operating in hot plasma—the conditions on the surface of the Sun. This needs to happen sooner or later and it is simply a matter of time, since the laws of physics allow it. When that happens, the Sun itself becomes a target for transformation into conscious material and its physical processes can become controllable. The Solar System may be the most developed planetary system in the galaxy, but this is not certain. The other systems would have had enough time to develop and some of them may have already outperformed the Solar System and be well into the process of transforming their stars into conscious matter. Moreover, there are stars at different stages of their life cycle around the galaxy, with different temperatures and chemical composition. This will create a range of conscious biochemistries which can transform into each other in the process of cooling or heating up of the stars. This would be a slow process, however, which would occur on the time scale of millions of years. What is important at this moment is that the possibility of consciousness in the plasma aggregate state will be achieved.

The planetary consciousness at low temperatures and the solar consciousness at high temperatures will be initially disconnected, communicating only via electromagnetic radiation. This would reflect the fact that there are two realms of physical matter existing in the planetary system—the interconnected planets some distance away from the Sun, and the Sun itself. The effort at that time would be to bridge that gap and populate the space between the planets and the Sun by conscious structures existing at gradually increasing temperatures and force of gravity. This will be a slow process taking place on a scale of hundreds of thousands of years, but when it is completed it will merge the planetary system in one whole and allow it to behave as a whole. Conscious processes then will become even slower and more spread out in the biomass, i.e., there is not going to be one central place, like the brain, where matter is organized in conscious structures, but most of the structures in the biomass will be adapted enough to each other in order to sustain a continuous conscious process. This process will be much more complex and more slowly evolving that the conscious process in a human brain, but it will be also simpler in some ways, since the outside world, the one it interacts with, will be more uniform than the world of human experience. This direction of development will set the stage for the transformation into a uniform, regularly structured and dispersed matter which will form the end state of the current epoch some billions of years into the future. So, the process of evolution of consciousness is not simply towards more and more complexly organized consciousness, but it also has a tendency for simplifying consciousness as the environment becomes transformed more and more successfully and becomes more and more structured. Consciousness, as we posited before, is based on the idea of a representation, so its complexity reflects the complexity of what it represents, i.e., the structures in the environment. One million years into the future represents a midpoint in the exponential evolution of consciousness in the universe and also an inflection point, after which conscious structures and their environment become more and more structured and more and more regular. Since it is difficult to imagine in detail the processes that far ahead in the future, we will make jumps of one order of magnitude from now on, skipping the intermediate stages.

1 million years into the future

That time point in the future, as already mentioned, marks the transition form the period of predominant creation of diversity of structures to the period of simplification and growing uniformity of structures. The conscious matter will be organized in a body of biomass with density of representations (i.e., adaptations of structures with respect to each other) comparable to that of the modern day human brain, so most of the living matter will be able to support conscious states in the form of resonant structures. Resonances will form and spread out as travelling waves in this biomass, interacting with each other in complex ways, and this activity will represent a new kind of a thought process, qualitatively different from that of the human thought. Let us recall that the human thought process has the basic form of a pulsating resonance, oscillating between a well-defined conscious state (highly integrated resonance), which is the state of maximum conscious awareness, and a loose resonance representing the transition between the well-defined conscious states, which is a state of minimum conscious awareness and goes largely unnoticed. The travelling waves of resonance in the biomass would represent a different form of organization of the conscious process, one where the well-defined conscious state is maintained continuously and it changes due to interaction with another travelling wave, so there will be no need for a dissipation mechanism which serves to transform the resonance and prevent it from getting "stuck" in one state. Still, most of the matter in the biomass will not be in a resonance at any given moment, so it will be in a non-conscious state, performing complex chemical transformations of materials, just like this happens in the body of a biological organism. The consciously organized matter will be orchestrating this process of transformation, with the goal to encompass more and more inanimate mass into conscious structures. The great diversity of living structures, existing throughout the whole spectrum of temperatures ranging from close to absolute zero in the interplanetary space to the 6,000 degrees on the surface of the Sun, will be an obstacle to the integration of all conscious processes into one whole, so there will be a great multitude of separate biomass structures with different biochemistries and individual conscious processes confined within them, but strongly communicating with each other, thus forming a highly integrated network of life forms, somewhat like the integration of neurons in the human brain. The direction of the evolution of this network will be towards greater integration and the formation of a continuous conscious process across the different biochemistries.

Similar development will occur throughout the galaxy in all colonized planetary systems. Since they all have different sizes and chemical composition, the details of the biochemistries and the composition of the conscious atomic structures will vary, but the trend of evolution towards a greater integration will be the same. The different planetary systems will also develop at various speeds and will be at different stages of their evolution. Some of them may still be in the stage of individual, freely moving life forms, like planet Earth today, and in others the life forms would have coalesced into one conscious biomass as in the Solar System. The different planetary systems will be in constant communication with each other, since their goal will be an even greater unification of the conscious processes into one giant galactic whole. This, however, would require a relatively uniform density of matter throughout space, and in even the most advanced in evolutionary terms planetary systems, the mass distribution will still be greatly non-uniform, with the star in the middle making up close to 99% of the total mass. The interplanetary space will also have non-uniform distribution of mass, with most of the planets still representing concentrations of mass orders of magnitude more dense than the space around them. This, however, will be gradually changing and more and more mass will be moved from the planets and the star into interplanetary space, where most of it will serve the purpose of capturing energy from the star to power the transformation process of inert materials into biomass. This movement of mass will require high degree of awareness in the biomass of the gravitational forces, in order to prevent gravitational collapse. In other words, the spreading out of mass into empty space goes against the fundamental law of increase of entropy and it requires intelligence in order to counteract this tendency and create greater order in the material structures, reflected in the more uniform distribution of matter. This process, however, is a very slow one, limited by the speed of light, and it will take millions of years to eject a significant proportion of the planetary material into interplanetary space. This task will be accomplished by the time period of 10 million years into the future that we will consider next.

10 million years into the future

At this time point in the future the main process under way in the Solar System will be the ejection of mass from the planetary bodies into interplanetary space, which will be already sparsely populated by low-temperature intelligent life forms. Relying only on energy from the sunlight falling on the Earth's surface it would take on the order of 1 billion years to eject all of its mass into interplanetary space, but the future life forms would have mastered also the direct conversion of mass into energy in thermonuclear reactions, as well as the harvesting of energy from sunlight in interplanetary space, so they will have more energy at their disposal than what is captured on the planet's surface. Their energy budget could easily be one to two orders of magnitude higher than the energy captured by Earth today, so the task of dismantling a planet and ejecting its material into interplanetary space could be accomplished on the order of tens or hundreds of millions of years. So, 10 million years into the future, most planets in the Solar System will be melting down and the interplanetary disk formed by the plane of the orbits of most planets will be filled by a sequence of rings of material structures endowed with intelligence. The biochemistry and the conscious processes in these rings will differ, depending on the distance from the Sun, which determines the temperature in that region of space. The surface of the Sun itself will be covered by a network of intelligent structures existing at high temperatures, and they will communicate with the intelligent life forms in orbit around the Sun in order to control the transfer of material from the Sun to them. The dissolution process of the planets will be accomplished by the next time period under consideration—100 million years into the future—and the main focus will switch on ejection of matter from the Sun with the help of the intelligent structures existing there.

The matter in orbit around the Sun will have lower density than the material structures existing earlier on the planetary surfaces, and correspondingly the conscious processes in the less dense structures will be slower, but also more regular. In the dense matter structures there is greater variety of material structures and physical phenomena associated with them, while in rarified matter the interactions between particles are more uniform and consequently the conscious processes will be more uniform. The aggregate state of matter in interplanetary space is not going to be between liquid and solid, as in the biomass on Earth, but it will have also properties of a gas in addition to those of liquid and solid, making the travelling waves of resonant coherence more uniform and spread out, leading to something like standing waves of resonance, due to the weaker coupling between the particles participating in a resonant structure. This would be a novel type of consciousness, one which is not possible in dense matter. This consciousness would be of the type of visual consciousness, since it arises from interactions of many informationally distinct resonances simultaneously, just like the multiple receptors of the retina interact simultaneously to form a two-dimensional signal, and unlike the conscious experience of sound, which arises from one-dimensional temporal sequence of stimuli. The interplanetary conscious matter may still retain some faculty of processing one-dimensional signals and the associated conscious experience of sound, but it will play a minor role, probably even less than in the human brain, since there will be very little useful information in that signal. The main conscious process will be vision-like, with one conscious visual structure interacting directly with another one (something like the way two brains would interact if they could be telepathically connected with each other), forming a four-dimensional mental realm (one two-dimensional visual signal interacting with another two-dimensional visual signal). This new mental realm, which will be an epiphenomenon, emergent from the interactions of particles existing and behaving in a three-dimensional universe, will become the scene for a new type of physics and new epiphenomenal space and matter, where a particle is effectively the resonant structure formed by particles existing in the three-dimensional universe and the interactions between these epiphenomenal particles follow different rules than those between the three-dimensional particles, effectively producing higher-dimensional space. In other words, this will be the birth of a four-dimensional universe with its own space and time arising from the new rules of interaction of the (conscious) material entities formed out of the material entities existing in the earlier, lower-dimensional universe. Note that the old three-dimensional universe does not exist anymore in that region where the new four-dimensional universe is created. The particles which interacted according to the physical laws of the three-dimensional universe until now become trapped in resonant structures and start to interact according to the new physical laws, effectively replacing the earlier interactions, meaning that the earlier three-dimensional universe has disappeared in that region of space and has been replaced by a four-dimensional universe. From now on, the four-dimensional universe begins to expand into the three-dimensional universe by virtue of incorporating more and more material structures into its novel material structures, transforming them from particles engaged in the old types of interactions into particles engaged in the new, four-dimensional types of interactions. That, however, can occur only if matter is spread out uniformly in space and that is why the process of ejecting mass from the star and the planets in the planetary systems is of vital importance.

So, 10 million years into the future the conscious structures will be advanced enough to engender the birth of four-dimensional structures in the Solar System and possibly in some other planetary systems in the galaxy, effectively producing a new Big Bang for the four-dimensional universe, but the rest of the three-dimensional universe (which is actually the vast majority of its mass) will be still lifeless and functioning according to the laws of physics familiar to us, so the colonization of the distant galaxies must continue. Within 10 million years it would be possible to reach the so called Local Group of galaxies by colonization missions travelling near the speed of light, and their transformation according to the pattern already known to us from the evolution of the Solar System would begin.

Beyond 10 million years into the future

The general process of transformation dominating the universe beyond the moment of birth of four-dimensional space will be the incorporation of old three-dimensional material structures into the new type of four-dimensional (or maybe even higher-dimensional) resonant structures. It is hard to judge from the current level of knowledge when exactly this transformation will start. It could be after 10 million years but it could also be after 10 billion years, or maybe even after 100 billion years, which would be enough time for the whole visible universe to be colonized and transformed into conscious material structures. Whatever the case may be, there seems to be only one direction for the evolution of material structures in the universe, namely, the expansion of consciousness, resulting in the engulfment of non-conscious material structures and their reorganization into conscious structures. This seems to be the most fundamental process governing the evolution of the universe and the one responsible for the phenomena of the Big Bang and subsequent expansion of space and maturing of material structures inside it according to the laws of physics pertinent to these material structures. This scenario seems to be repeating itself, forming distinct epochs in the evolution of the universe, each subsequent one involving more complex relations between material particles leading to higher dimensionality of the epiphenomenal space engendered by these interactions and different laws of physics. Currently we live in a three-dimensional universe, which suggests that there have been one or two epochs before our Big Bang with material particles and possibly conscious beings existing in lower-dimensional space (one or two dimensions), and the apparent success of string theory to explain at least partially the physics of elementary particles may be an echo from these earlier epochs, as we noted earlier. The strings in string theory are one-dimensional objects and their mathematical properties are a good match, although not a complete match, for the standard mathematical description of the laws of physics governing the interaction of elementary particles in three-dimensional space. Still, the mathematics of strings incorporates some unresolved problems which prevent it from being a viable description of the laws of physics at a more fundamental level than the experimentally verified standard model of elementary particle physics, but it has engendered lots of hope since it seems to work well in resolving some partial problems necessary for creating a Grand Unified Theory of particle physics. The fact that string theory manages to some degree, but not completely, to provide a fundamental theory of physics corroborates the ideas proposed in the previous paragraphs of a cyclical evolution of the universe through subsequent epochs of increasing complexity of interactions and dimensionality of space. If that is indeed the case, some of the fundamental character of the interactions will be preserved from the earlier epochs, hence the partial success of string theory, but the interactions in the subsequent epoch will necessarily be more complex in some respects, hence the inability of string theory to explain all the physics of the current epoch.

The Representational Picture of the World
(How to Imagine the Universe)

We started our exploration of the concept of matter with the classical picture of particles moving and interacting in empty space. Now that we have built up our understanding of matter, consciousness and the universe as made up of representations, we can make a contrast between the classical picture of how most people imagine matter and our new picture based on the notion of representation.

Modern-day physicists have moved a bit further from the classical picture of tiny balls made up of hard stuff flying in empty space and colliding, but they still cling on to the idea of particles being fundamentally identical in their structure and behavior (e.g., all electrons are the same) and the deviations from the precise values of their properties (such as mass, energy, position in space relative to other particles, etc.) are regarded as a result of a fundamental indeterminacy due to random quantum variations (or noise). Thus, the view of matter, and therefore the cosmological theories of the universe as the totality of all matter, in modern physics is based on the assumption of perfect sameness of all particles and unavoidable fluctuations of their properties which are a consequence of the quantized nature of matter. However, modern physics does not have an answer as to why the particles should be identical and quantized—this is considered just a firmly established empirical fact.

Our representational picture of matter, on the other hand, is based on very different postulates regarding the identity of the particles and the source of the variations in their properties. If particles are indeed representations in their most intimate nature, then all the variations in their properties and their states should be regarded as their true character. In other words, there is no single prototype that stands as the 'true' character of the particle (in terms of the values of its properties, such as mass, rest energy, direction of the spin vector, etc.), which is being constantly perturbed by quantum fluctuations, but this perceived prototype is just a highly predictable average of the individual characters of many particles and every individual deviation from the average is a 'true' character of the particle in its own right.

This way of thinking about the identity of an object from a class of similar objects with slightly varying properties is familiar to researchers in other fields of science, such as psychology, biology, etc, who have to deal with variations in macroscopic objects of the same kind. For example, if we take hundreds of pictures of faces of people from the same race, gender and age, when we compute an average face based on the most straightforward averaging algorithm, we will obtain with a very high probability one and the same average face no matter which sample of faces we take from the whole population. In this case we are not led to the conclusion that there is one 'true' standard face, and all actual faces of people are random deviations from the standard, but we think of the individual faces as real and the average as epiphenomenal. Within our representational picture of matter the same reasoning should apply to particles—their individual states are real, because each representation reflects the peculiar circumstances in its own environment, and their averages are epiphenomenal. The fact that the averages are highly reproducible in similar circumstances (e.g., temperature, pressure, speed of movement, etc.), while the individual deviations are not, should not lead us to believe that the averages are real, but to the conclusion that there is a degree of sameness in the representations which does get expressed in the same way through the particles' behavior when we arrange for the measurements to take place in similar circumstances.

To sum it up, in the representational picture of matter both the sameness component and the difference component of the representations play an equally important role—they are both equally 'real' and define the character and behavior of a particle together. This is in contrast to the conventional picture of matter, where the sameness of all particles of the same type is considered 'real' and the deviations in particles' properties are considered 'noise'. This difference between the two views has profound implications on how we imagine the universe and how we reason about it in terms of our basic ontology.

In the modern cosmological picture of the universe the particles of matter are considered independent entities, whose number and movements in space, and therefore their interactions as well, are determined by the initial conditions in the moments briefly after the Big Bang. If there were no quantum indeterminacy, the movements of particles would be completely deterministic and the universe could be conceived as a giant mechanism working like a clock, but the random quantum fluctuations add noise to the system of moving and interacting particles and render its evolution predictable only in probabilistic terms. This picture is again in contrast to the way we imagine the totality of all particles of matter within the representational conceptual framework.

In the representational picture there is no randomness and the movements and interactions of the particles, viewed as representational complexes, are completely determined by their environments, i.e., the set of other representational complexes with which they are in a representational relation. Thus, the behavior of each particle has its causes, which can be traced back to the very origin of time and space in the primordial Big Bang, but that does not mean that it is possible to *know* everything about the universe and to simulate it perfectly—in the representational framework this is as impossible as in the standard cosmological picture which postulates the existence of a fundamental randomness.

The reason for this is that in a system of representations with varying complexity the more complex representational constituents can represent the more simple ones, but the opposite is not the case, so the most complex representational complexes remain unrepresented, and therefore unknown and non-simulatable, by any other structure. This is especially true in light of the law of accumulation of representations, leading to an evolutionary process producing increasingly complex representational structures. As those structures become more and more complex, as is the case with brains, culminating in the complexity of the human brain, they become more and more difficult to understand and to simulate, and the effort to achieve that understanding makes them in fact even more complex. In other words, as our knowledge of the brain increases, so does our technology and our ability to manipulate the brain, which in fact leads to an increased complexity of the brain-environment system. It is possible to envisage our theoretical knowledge advancing faster then the technological augmenting of the brain, but even in this case we may achieve full knowledge of the brain only in principle—for an idealization of the brain—but not for every individually existing brain and its particular state at any moment in time. That means that there will always be unrepresented parts of physical reality, which are too complex to be understood by a conscious entity.

The logical impossibility of complete representation, and therefore knowledge, of the totality of all representations is an important conclusion which we arrive at when thinking about the universe as a whole—the totality of all representations. It is not a necessary conclusion, though, if we just think of matter, or systems of particles, in an abstract way, or if we imagine the universe as having an infinite extension in space. Then it is always possible to imagine a more complex being which would be able to acquire perfect knowledge of any finite system of representations, no matter how complex it is. However, in order to do that this entity itself must possess an even higher degree of complexity, and only the finite nature of the totality of all representational structures (meaning finite extension of space and finite granularity of the representational elements) ensures that there is a limit to how complex a representational structure can grow and therefore that the most complex structure will remain unrepresented.

So, in general we can say that any infinities—either in terms of increasingly large structures or increasingly small structures—lead to logical absurdities and to the possibility for anything to happen or to exist in a universe made up of constituent units with finite size. However, infinities are well established in the conventional picture of the world described by the different branches of physics, since the mathematical apparatus used for describing the laws governing the behavior of matter is built up on the notion of infinity. For example, a number is represented as an infinitely small (zero-dimensional) point on the number line, and physical particles too are regarded as point-like entities in order to make the mathematical apparatus describing them work. The universe itself is considered having infinite extension, leading to different scenarios of multiple universes existing simultaneously. The string theory deals with entities—strings—which have finite extension in only one dimension, but are infinitely small in the other two, when situated in three-dimensional space. All these conceptual devices are very useful for the purpose of making mathematical models of physical reality, but they are inadequate for understanding what matter *really* is and building up an imagination of how it works—for that purpose the idea of representation without any infinities seems to be more appropriate.

These views on the source of randomness and the necessary existence of unrepresented material structures in the universe have consequences also for the question of determinism and free will, which we discussed in relation to human consciousness in Chapter 2. There we posited that free will stems from the constant creation of new representations in the fundamental process of accumulation of representations, which entails an unrepresented top of the hierarchy of representational relations between material structures. The finite size of the universe, as we saw, ensures that this hierarchy is limited and indeed has a top. So, we need to regard human consciousness and its future offshoots as a constantly moving target for our knowledge contained in our conscious material structures, which gets more and more complex the more knowledge we accumulate, thus continuously escaping full understanding. This, in fact, is the source of free will in the universe—not just human consciousness, but any unrepresented material structure would appear as acting on its own and affecting the other material structures in unpredictable ways.

The process of accumulation of representations cannot lead to complete representation of all material structures, because that would mean complete sameness and therefore collapse of all representations to a single, immutable one, i.e., the end of time and space and the death of the universe. This too seems a logical impossibility, so we need to imagine the material structures in the universe as achieving a limit of the degree of sameness and starting to build up a higher-order type of representations with a high degree of difference and a small degree of sameness, just as in our discussion of the dimensionality of the current stage of the existence of the universe and the successive stages separated by Big Bang-type of phase transitions.

Thus, our picture of the universe becomes one of a finite space and a finite number of simple representations (particles) which transform, and in doing this change their representational configurations, which looks like interactions between pairs of particles, leading to an ever increasing accumulation of representations which has no logical end. So, although the set of representations has a logically simplest state, which looks like the origin of space and time, it does not need to have an end in a logically most complex state, rendering the time evolution of the universe as an infinite process, and this is the only infinity in the representational picture of the universe. Logically, this picture is the most parsimonious, and therefore also the most consistent, synthesis of our current knowledge of the world in a single, integrated conceptual framework.

Chapter 4: What Does It All Mean

What is the Meaning of Life

After building our understanding of inanimate matter and the conscious organization of matter and the ensuing consequences for the evolution of the totality of everything, which we call the universe, we are in a good position to draw also some conclusions regarding our contemporary knowledge of the world, which is more concerned with more mundane and immediate issues from everyday life, such as money, health, social interactions, etc. Although less fundamental than the knowledge that we built up in the preceding chapters, the understanding of such questions is of great importance too, since it represents the current state of affairs of the consciously organized matter in the world. The fact that most people are more interested in those less fundamental issues and more knowledgeable about them, in contrast to the more fundamental knowledge of matter and consciousness, is simply a fact of life, just the way things are, and it has to be accepted as a given and taken into consideration.

Probably the most immediate conclusion we can draw from our picture of the world as a system of interacting representations with regard to our orthodox understanding of the world concerns the perennial philosophical question of the meaning of human life, or what is the right thing to do; the right course of action and the right understanding of that action. Traditionally, this question has had a sizeable moral dimension and has been rooted in the religious understanding of the world, but it can also be conceived as a more pragmatic question of the role of humans in the world and the purpose of their existence.

Having reduced our understanding of matter and consciousness to the picture of interacting representations, we are well prepared to answer the pragmatic side of the question of the meaning of human life. As we saw, a human being is just an aggregation of particles organized in living and conscious types of structures, understood in an abstract sense as densely accumulated representations. In fact, human brains are regarded as the most complex structures in the universe, which in our terminology translates to the most densely packed complex representations, and therefore their role in a universe made up of representations is to guide the transformation process of all representational structures. In other words, humans are at the forefront of the transformation process leading from the three-dimensional organization of matter to the four-dimensional one that will engender a new universe.

So, this is how one can define the role of humans, and therefore the meaning of human life, within the conceptual framework of the picture of the world based on the notion of representations. In more mundane terms this can be translated into statements of the kind of "humans are destined to transform the world and bring it to a new level of organization of matter using their conscious abilities, and the meaning of one's life is to promote this process". Even if we do not adhere to the hypothesis of the representational nature of matter giving rise to an evolutionary process of accumulation of representations and epiphenomenal physical space with a specific dimensionality, we can still think of humans as destined to transform the world and define the meaning of one's life along those lines.

Can we find support for our proposition regarding the meaning of human life in the current state of affairs in our world? It seems a fair statement to say that modern humans are preoccupied mostly with making their lives easier and improving their wellbeing. This motivation is underlying most of the reasoning processes in our political, social and everyday activities, reflected in the topics of interest most prevalent in the media. But what does it mean to improve our wellbeing and to make our lives easier? In most general terms, we can think of it as a process of adapting our physical environment to suit our needs. Physically, this is a process of taking raw materials from the surface of the Earth and transforming them through different mechanical and chemical manipulations into finished goods with complex structure which serve some useful purpose. The majority of people on Earth are involved directly in this process (i.e., in the agriculture and industry sectors of the economy), and the rest take up the role of coordinating and managing this process (the services sector of the economy, apart from those in non-productive sectors such as entertainment, sports, arts, etc.).

In terms of representations, the manufacture of goods is in fact a process of transformation of particle structures forming the raw materials into particle structures forming the finished goods that are arranged in such a way that they are more adapted to, and therefore more representative of, the human needs, i.e., the functioning of the human body and human thought. So, viewed in this way, the process of manufacturing of goods can actually be understood as (and reduced to) a process of rearrangements of particle structures leading to a higher degree of representational adaptation between structures in the environment (mostly inanimate, but we manage also the biological organisms and adapt them to suit our needs) and the structures that constitute us, humans, who are the driving force of this process.

It should be noted that a substantial part of human activity concerns transformations of particle structures belonging to the conscious organization of matter, i.e., the thought processes of other people. In economically developed societies this is the prevalent type of activity, where people are engaged not in the transformation of raw materials into finished goods, but in the transformation of other people's thought processes via different physical modes of communication (computers, printed materials, electronic media, etc.). This is essentially true for all work done in an office and not on a farm or in a factory, including activities as diverse as accounting, artistic design, architecture, business management, politics, science, etc. All these activities in some way reflect or aid the production of goods, but do not directly partake in it.

So, human activity, which is for the most part directed at transforming particle structures in ways which make them more adapted to the functioning of the human body and mind, can also be viewed from an outside perspective as part of the general process of accumulation of representations governing the evolution of material structures in the universe. The human-induced creation of new representations has two major components—transformations of particle structures belonging to the non-conscious organization of matter (industrial manufacture of goods) and transformations of consciously organized particle structures (services and creation of new knowledge). The latter part is activity which remains more often unnoticed, since it concerns the organization of matter inside the human brain, and it is not obvious within the prevailing ontological picture of the world that it too can be reduced to a physical process of rearrangements of particle structures.

That inability of the conventional conceptual framework, used to understand the world, of visualizing the mental processes inside the brain and how they relate to physical processes outside the human body actually underlies also the difficulty in resolving such a question like that of the meaning of human life. This question seems difficult if we do not have a good idea of what consciousness is and how the workings of the brain relate to the inanimate objects in the environment, but using our framework of representations and the understanding of consciousness and physical processes within that framework this question becomes easy to answer.

The conceptual framework of the world as made up of representations has the potential to resolve not only the long-standing question of the meaning of human life, but also many other, if not all, traditional philosophical questions, given that we have built up an understanding of the workings of the conscious and inanimate structures in terms of representations. In the following section we will try to outline the main discrepancies between the traditional way of thinking and our picture of the world based on the notion of representations, and suggest ways to resolve them, i.e., prescribe general rules that need to be observed when reasoning and formulating new knowledge. In this way, the picture of the world based on representations, if adopted by a number of people, can lead to a more adequate understanding of the world, which is complete at some level of knowledge and therefore can answer any well-formulated question, which in turn would lead to a greater degree of agreement and more successful action within this group of people. In a sense, the ability to understand the world through the framework of representations promotes the fundamental process at work in the universe that we formulated within that framework, the one that is also key to the answer of the question of the meaning of human life.

The Role of Knowledge (How to Think Correctly)

The term knowledge within our framework of representations refers to arrangements of particles within one's brain and how well the representations they form reflect arrangements of particles outside of it (including ones within other brains). The awareness of the workings of the human brain and the physical substrate of knowledge, as defined in the previous sentence, makes it easy to imagine the world in purely physical terms and consequently to resolve any well formulated question. Most of the difficulties in the traditional understanding of the world and the seemingly irresolvable questions that arise are in fact due to the lack of such awareness stemming from a lack of understanding of what consciousness is. In this way, traditional reasoning can only establish relative, or conditional, values and judgments, but not absolute ones, since it often reaches irresolvable questions along the chain of reasoning. In order to be able to resolve any question, and as a consequence to be able to think correctly within a given conceptual framework, one needs to be able to reduce any question to a formulation involving a single concept. In the conventional philosophical paradigm anything can be reduced to the concept of matter, which captures very well the workings of inanimate material structures, but there remains a big mystery pertaining to the workings of living organisms and especially the human brain, captured by the philosophical notion of consciousness. This lack of understanding of the workings of consciously organized matter and its relation to inanimate matter is what prevents traditional reasoning from achieving completeness and the ability to resolve any question.

So, is it really possible to reduce every question or proposition to a formulation involving a single concept? This is in fact an idea with a long-standing tradition in mathematics and in physics. Mathematicians strive to derive all mathematical propositions from a minimal set of axioms, while physicists are on a quest for a single equation from which all other equations in physics can be derived. These are both examples of logical (i.e., reasoning) systems with a hierarchical structure, where all the different constructs can in principle be reduced to a small set of concepts or even a single concept.

The concept of representation may indeed be able to play the role of a single most fundamental concept for a comprehensive conceptual system capturing all human knowledge, but the practical question of whether it is feasible to imagine everything as different manifestations of interacting representations still remains. Intuitively, it seems hard to imagine for a person not used to thinking in terms of representations how concepts such as love, impatience, friendliness, etc., or daily worries such as how to fix a broken car, or how to cook a certain meal, can be thought of only in terms of representations; however, this too should be possible if indeed the world is best understood as a (rather large) collection of representations.

The difficulty, as we said, stems from the fact that conventional reasoning has not developed the habit of imagining what goes on inside the brains of people in physical terms, but the workings of the human mind are instead imagined in an abstract, non-physical way. The abstract concepts are formulated independently of each other, and as such they do not reduce to a single underlying concept (e.g., love, friendliness and impatience are generally not thought of as manifestations of some more basic property of the mind, such as consciousness). This leads to the possibility of conflicting views, due to the variability of the conceptual structures made up of abstract concepts for different people, and therefore to irresolvable questions due to the lack of an arbitration mechanism based on reducibility of the abstract concepts to some common, more fundamental concepts.

So, we come to the conclusion that abstract concepts are actually detrimental to the correct understanding of the world. They may be thought of as temporary tools which one can use in the lack of proper ones, but which should be discarded once one finds the right conceptual tools from which one can build a complete and consistent picture of reality. This is one of the main lessons we can learn from the picture of the world based on representations that we developed in the preceding chapters—it makes a good practice to think of everything only in physical terms and to avoid abstractions altogether, or at least to use them temporarily, until one learns to translate them to the workings of material structures.

This lesson is especially relevant when thinking about the main topics of interest for the wider public—the economic, political, and social life, and the decision-making process that is at the heart of it. Currently, there is a large variety of opinions on most issues in these subjects, and since most of the knowledge pertaining to these subjects is constituted in abstract concepts that are not reducible to a single more fundamental concept, the large variety of views on how things work (or should work) leads to many controversies and to unresolved questions. On the other hand, in principle the truth is only one, meaning that there is only one most optimal solution to any issue, and only one most optimal decision or course of action. So, if everyone involved in this decision-making process was able to think about the issues in purely physical terms, i.e., to reduce all abstract concepts to concrete, physical ones, that would make it possible to eventually settle any debate by achieving agreement on the views of how things work and what is the optimal course of action.

That might very well turn out to be not only the sufficient, but also the necessary condition for reaching agreement, since it seems very unlikely that the large variety of abstract concepts existing inside people's minds would ever converge to a single, coherent logical system of reasoning. Of course, in order to reach agreement all participants must be consciously making the effort to reduce all their knowledge to different manifestations of a single concept, otherwise there is no guarantee that such a reduction will be achieved even if it is possible in principle.

So, agreement is important for avoiding conflict and for harmonious social life and the progress of civilization, but this should not be just any kind of agreement, it must be agreement on the right worldview, i.e., on the most truthful fundamental understanding of reality. For the moment the best conceptual system for understanding reality is that of science, since it is based entirely on empirical evidence and rational arguments. There are other competing fundamental explanations of reality, mostly grounded in religious traditions, but they are based on beliefs rather than rational understanding and are therefore incorrect representations of reality, leaving science as the only contender for best explanatory system. The trouble with science is that it still has not been able to explain major empirical phenomena, such as consciousness, and until it does so the alternative worldviews can hold their ground and agreement on a single system of knowledge would be out of reach. In other words, science must achieve the creation of a complete picture of the world at some level of detail, and only then complete agreement and smoothly functioning social and personal life will be possible.

In that process the ability to think correctly is indispensable, and at the heart of it lies the ability to reduce all concepts to the workings of a single, most fundamental concept such as that of representation. Doing that would produce a body of knowledge, i.e., patterns of arrangements of particle structures inside all human brains, which is in some sense the most accurate representation of external reality (including those patterns in other people's brains, since they are part of the external reality for a given person). That body of knowledge as a physical structure will be the most advanced entity along the trajectory of the evolutionary process of matter in the universe, and therefore will be the driving force of this process. In fact, the collection of all consciously organized matter inside human brains at the present moment is at the forefront of the evolutionary process too, but due to the diversity of human knowledge, reflected by the fact that there is no agreement on the fundamentals, the consciously organized matter does not drive the evolutionary process smoothly in the right direction but pulls it in all sorts of different directions and dead ends, and only on average this erratic motion advances along the right direction.

The diversity of knowledge is realized by a diversity of representations in the human brains with different degree of veridicality, i.e., of correspondence with the external reality. In this way every representation can be assigned a measure of veridicality corresponding to the degree of truthfulness of this representation with respect to the state of affairs it represents. If the representation is highly truthful, then we should consider this representation correct, and if it has low veridicality, then we should consider it wrong. In the next section we will examine the consequences for the body of knowledge of the existence of these two types of representations.

What is Right and What is Wrong

Right and wrong are concepts, i.e., well-defined mental states, associated with specific feelings, or phenomenal experiences, related to the situational content of what is being judged as right and wrong. More specifically, the words "right" and "wrong" are adjectives, meaning that they qualify other words that play the role of nouns or factual statements. In that way, those qualifiers denote a property of the representational relation between concepts denoting a mental state of affairs (our beliefs) and concepts denoting the state of affairs in the external world (the content of the beliefs). That property is the veridicality, or correspondence, of the current mental picture with the picture derived from previous experience.

So, crudely, we can say that the concept of "wrong" is based on a mismatch of two representations formed simultaneously in the brain, one based on current experience or reasoning, and the other based on past experience, while the concept of "right" is based on a match between two representations of this kind. As we saw in chapter 2, the interactions between representations are the physical substrate of perceptions and emotions, so the distinct physical characteristics of the synchronous evolution of shared entangled states, which supposedly realize the concept of "right", and asynchronous evolution in the case of "wrong", are sufficiently distinct and persistent in order to form their well-defined mental states and therefore to be recognized as distinct concepts.

They are special concepts, however, since they correspond to the most basic properties of a representation—the degrees of similarity, or match, and difference, or mismatch. Because of that it turns out that our judgments of what is right and what is wrong are much more important that other judgments, like those of what is pleasant (although it is correlated with what is right and therefore it is important too), what is tasty, what is boring, etc. In terms of the evolutionary process of matter in the universe we can say that the judgments of what is right and what is wrong play the most substantial role in the decision making process guiding people's actions, and therefore affect the most the direction of the evolutionary process, which we can think of as the totality of all actions of all people in a given period of time.

The contrasting concepts of "right" and "wrong" are closely related to another pair of contrasting concepts—of "good" and "bad" (or "evil"), with the latter being more abstract and typically not considered reducible to the physical level. "Good" and "bad" are even more emotionally charged than "right" and "wrong" and are not necessarily justified as judgments through rational arguments, but are allowed to be sometimes subjective opinions. Nevertheless, according to our proposition in the previous section that every concept can be reduced to the physical level, they can be reduced too, and this can be done via the concepts of "right" and "wrong", and more specifically through the fundamental notions of matching and mismatching representations.

So, what is "good" and "bad" in a picture of the world made up entirely of representations? As concepts they are well defined shared entangled states that interact with other well defined shared entangled states. When these interactions are with compatible states, i.e., they form to a large degree matching representations, they are associated with a distinct positive feeling which accompanies the instantiation of the concept of "good", and vice versa for the concept of "bad". These concepts are in addition situated in a conceptual system based on experience and on the specific functioning of the human body. This means that they depend on the knowledge of a particular person, i.e., the entire set of concepts and their relations, and on the idiosyncrasies of the emotional and rational makeup of one's personality (i.e., again, the specifics of how the different well defined shared entangled states relate to each other). In this way, the common property of the concepts of "good" and "bad" instantiated on different occasions in different people is that they are associated with matching and mismatching representational shared entangled states, but there are individual differences arising from the specific makeup of people's knowledge and emotional systems.

It is those differences that lead to disagreement and to conflict, but this diversity of conceptual structures is also what allows for creation of new knowledge and for improvements of the conceptual systems. This situation is reflected in the large diversity of opinions existing nowadays on many issues with a moral aspect to them, involving judgments of good and bad and of right and wrong. The differences in the moral judgments people make are based on differences in their knowledge systems, and since no one has a complete picture of the world, they result from the different ways these knowledge systems are incomplete. So, "good" and "bad" and "right" and "wrong" are rationally constructed, but the different outcomes of judgments are due to different missing pieces of knowledge.

Since the judgments are rationally constructed, there is usually some valid justification for any argument, even in the case of conflicting opinions. In such cases the choice is typically between two good or two bad options and the judgment that informs the decision is about the relative magnitudes of the two options. For example, the government of a country that needs to deal with overpopulation has to choose between two undesirable courses of action—either to force people to have fewer children or to invite more hardships in the future when it will need to feed a larger number of people. In such cases inevitably some people will tend to give more weight on one side of the argument for and against forceful limitation of the birth rate, and others will choose the other side. The differences are based on different lines of reasoning that can be traced down to different perception of what is a human being, how valuable is one's life, etc. In summary, the different lines of reasoning come from different conceptual systems, which we may also call worldviews, but given that the world is one integrated system, there is only one worldview which is most veridical, i.e., closest to reality, and therefore only one line of reasoning is the most appropriate one, while all others are suboptimal and therefore should be judged wrong. In this way, there is only one right decision and course of action, which in our example can be either limiting the birth rate or coping with a growing population, and the other option is objectively wrong.

So, we conclude that in principle there is always one right line of reasoning and all others are wrong. The problem with this proposition is that due to the incompleteness of our collective and individual knowledge it is impossible even in principle to know which is the right line of reasoning. This would be possible only if all people participating in the decision making process possess a complete conceptual system, meaning that it is a complete representation of reality at some level of detail (and therefore there are no unanswerable questions within this system), and moreover, they all possess more or less the same conceptual system, i.e., they agree on the fundamentals. Since this is not the case at the present moment, there is no way to decide with absolute certainty on any occasion what is objectively the right course of action.

The judgments about what is right and what is wrong are further complicated by the fact that they typically refer to future actions and events, which are merely potentialities but not actualities, with the added complication that the uncertainty in one's knowledge of the state of affairs at the present moment is compounded with the uncertainty in the future evolution of this state. Retroactive judgments about past events are typically easier and more conclusive, although there is uncertainty there too due to our unavoidably incomplete knowledge of the past. So, we can be more successful and more certain on average in our judgments about the single objectively right course of action in the past, but much less so for the future.

In summary, we can say that the judgments about right and wrong can be singled out as the most significant mental acts since they are the representational mental states that are closest to the most intimate nature of matter—the matching and mismatching aspects of the representations. Although these are just feelings pertaining to the internal representational dynamics of the cognitive structure of mental states in the brain, by virtue of the representational relation between this cognitive structure and the external environment, i.e., the material structures outside the brain, these feelings play crucial role in determining one's actions and ultimately the totality of all actions by all people, which sets the direction of the evolutionary process of matter in the universe.

Because of this crucial role of our judgments about right and wrong (and also good and bad) mistakes are costly and being close to the truth is very important. Mistakes, however, are ubiquitous and unavoidable, given our partial knowledge of the world and the large diversity of opinions. The good news is that on average we are moving closer to the truth and progressing towards arrangements of material structures that are more adapted to each other and more advanced along the direction of the evolutionary process of matter. This means that the correct judgments outnumber the mistakes, and this is so because most of our actions and thoughts in our daily lives are routine; they are simply repetitions of well-rehearsed acts and thoughts in varying combinations with little novelty which could result in mistakes. Repetition and reinstantiation of earlier states is the hallmark of living systems, evident in phenomena ranging from the reproduction of organisms and their DNA to language, which is a stream of well-rehearsed mental states corresponding to the words accompanying the thought process. This repetition constitutes for the most part correct action, and the novelty introduced by novel combinations of old elements contains both improvements and mistakes (just like the mutations in DNA). Our judgments of right and wrong act as an agent of natural selection in the domain of ideas (or mental states), guiding the evolution of our collective cognitive system in the direction of more veridical representation of the world, i.e., closer to the truth.

Given our conclusion regarding free will, namely, that although the processes inside our brains are deterministic, they are not represented by any other material structures and are therefore unknowable, it means that we have personal responsibility for our judgments of right and wrong and for any mistakes we make. Our conscious efforts can change the outcomes of our judgments and therefore can affect the evolutionary process as a whole (although usually to a very minute degree). Still, for practical purposes we should think of ourselves as possessing free will and carrying responsibility for our actions and judgments.

Making correct judgments is based on having a veridical system of knowledge of the world, and therefore the primary responsibility of any person is to build such a system of knowledge, i.e., to know the truth as much as possible. This demands both extensive knowledge (quantity) and correct knowledge (quality). As we stated earlier, veridicality is best achieved by thinking only in physical terms and avoiding abstractions. In other words, our knowledge should be grounded in empirical observations and conclusions based on *a priori* assertions should be avoided, as is indeed the case with scientific knowledge. However, even scientific knowledge contains a great deal of abstractions, and therefore mistakes, due to the lack of a single most fundamental concept to which all knowledge can be reduced. Therefore, the natural direction of development of our collective knowledge in the near future is towards a more scientific way of thinking for the general public and towards physicality, away from abstractions, for scholars, especially for knowledge pertaining to the mind and communities of minds.

The current state of affairs of our collective knowledge and its projected developments in the near future will be the topic of discussion in the remaining sections.

The Present State of Human Thought

Imagining the world in physical terms means imagining arrangements of particles as representations of other arrangements of particles. As a rule, the particle structures making up human brains form representations of other particle structures, while the same is not true for particle structures belonging to inanimate matter and holds to a much lesser degree for particle structures making up living organisms but situated outside the brain. That is why our most important task in building up a veridical picture of the world in physical terms is to develop a realistic understanding of the representational content inside the actual human brains (or minds) that exist at the present moment. This means primarily categorizing the unique mental states (which never repeat exactly, as we saw) into a not very large number of groups and getting right the relative proportions of these groups in terms of both frequency and importance.

The crudest level of categorization is to divide the mental states in two groups—those pertaining to material structures outside the brain and those reflecting the mental states of other people (also in abstract terms). Most people nowadays are engaged most of their time in manipulations of physical objects in their environment, be it in the work place or at home. Even when people are engaged in a conversation and actively imagine other people's mental states, there is still a component of physicality in people's mental states, involving their perceptions of the physical environment, the visual image of the other person, etc. In general, imagination of non-physical entities, such as other people's mental states and abstractions, are more difficult (especially earlier in life) and therefore less frequent.

On the other hand, the manipulations of physical objects are more mundane and are judged less significant, while picturing other people's mental states and abstractions is perceived as more interesting and leaves deeper impressions in people's minds. Thus, we can say that although the first category of mental states—those pertaining to material structures outside the brain—dominates in terms of frequency, the second category—pertaining to other people's mental states—dominates in terms of importance.

There are individual differences, as well as cross-cultural differences, in the relative prevalence of each of these two categories of mental states. For example, people in economically less developed societies are engaged to a larger degree in manual labor and their mental lives involve lesser use of abstractions, so in comparison with people in the developed countries their conceptual systems represent a cruder reflection of the external world and the hierarchies of concepts in terms of levels of abstractions for them are more shallow. The lesser use of abstractions, however, is not a sign of higher veridicality, but rather a consequence of the smaller size of their conceptual network. Being smaller necessarily means that it is a cruder reflection of the real world and therefore less effective. The more sophisticated conceptual systems of people in more advanced societies mirror better the processes and structures of particles in the external world, including their larger-sized abstraction components, however, even though a significant proportion of the abstract conceptual networks fails to match closely reality, it is still the smaller portion of the abstraction component, while the larger part of it reflects reality close to veridicality.

The same is true for people with different social and educational standing within a society. There are large individual differences, with people with better standing having more sophisticated conceptual knowledge systems with larger abstraction components than people with lower standing, who tend to participate more often in manual labor activities. In this way the level of personal and communal economic development correlates with the level of development of people's conceptual knowledge systems and the way it reflects reality. This observation reflects our earlier conclusion regarding the meaning of human life as the transformation of our environment in line with the evolutionary process of matter in the universe. Since economic activity reflects the physical transformation of the surface of the Earth to suit better our needs, its correlation with the development of the conceptual knowledge systems reflects their representational nature, i.e., the fact that they also develop in line with the fundamental evolutionary process of accumulation of representations.

If we imagine the conceptual knowledge systems as entities in abstract state space, as we commonly did in the previous chapter, then the distinction we draw between less advanced and more advanced knowledge systems would become a distinction between a smaller and less active structure of particles in abstract state space and a larger and more active one. The difference in the levels of activity corresponds to the observation that people in more developed societies are better learners and incorporate new facts in their knowledge systems with greater ease, compared with people in less developed societies, although once those people acquire the desire and motivation to learn they can easily catch up with their more able counterparts. In purely physical terms, the larger size of the particle structure in abstract state space is what makes the level of activity higher, since each particle (i.e., concept) is exposed to more influences from other particles and therefore is more actively transforming, leading to increased probability of transformation of complex particle structures (i.e., conceptual schemas, or knowledge episodes). This is akin to the process of heating up of a body of particles in our three-dimensional universe as it accumulates more mass and compresses under the force of gravity. The more frequent readjustments in the abstract conceptual structure is what makes it more veridical, i.e., better representing the outside reality in real physical space (represented by particles in the surroundings of the conceptual structure in abstract state space).

The distinction of more advanced versus less advanced knowledge systems has one more aspect—that of the differences in the truthfulness of people's beliefs and opinions. As we mentioned, people with more advanced knowledge systems typically hold beliefs and opinions that are closer to the objective state of affairs compared to people with less advanced knowledge systems. This holds true both for comparisons in parallel between people with similar social roles and status belonging to social groups with different degrees of development, as well as for longitudinal comparisons between different stages in one's life, given that for the most part of one's life one's knowledge system develops continuously towards a more advanced state. In all these cases the greater sophistication of the knowledge system entails more veridical representation of the external state of affairs and therefore greater degree of truthfulness of one's beliefs and opinions.

To put it bluntly, this means that people belonging to social groups at a more advanced stage of development are more often correct than people from social groups at a less advanced stage of development, who in turn make more mistakes in their judgments, translating into more mistakes in their work and domestic activities, and consequently higher degrees of conflict and problems in their life experiences.

The different degrees of truthfulness of people's beliefs and opinions are reflected in the coarseness of the conceptual structures and schemas that they employ in making sense of the world around them. People with less advanced knowledge systems typically employ coarser conceptual constructs in their explanations of the situations and events they witness in their lives, leading to more radical views and opinions that follow simpler logic, while people with more advanced knowledge systems are able to construct more sophisticated explanations and conceptual schemes, giving them better chance to match the objective reality more closely. In other words, people in more advanced societies have more graded beliefs and opinions, while people in less advanced societies have simpler and more radical views.

It should be noted that the greater sophistication and its corresponding finer granularity of the representational relation between the mental construct and the represented state of affairs in the world is not a guarantee for higher veridicality and better understanding; it is so only on average, but the higher degree of sophistication entails also a higher level of abstraction and that in turn entails a higher chance of misrepresenting the real state of affairs, i.e., of forming a false belief. This phenomenon is one of the defining characteristics of modern-day advanced societies—the greater sophistication and higher volume of knowledge lead not only to a better capability to transform the environment, expressed in a higher level of economic development, but also in a greater intensity of the debates on all aspects of life and greater need to deal with more and more complex problems at the workplace and in the private life. These are all characteristics of an increasing complexity of a system, signifying progress along the trajectory of the main evolutionary process of accumulation of representations.

The growth in complexity means also growth of the number of topics making up the system of knowledge of all humanity and even growth (although to a much lesser degree) of the number of linguistic units (i.e., words) that people use. Our knowledge is captured for the most part in language constructs reflecting sequences of mental states that originated in the brains of one or more people who produced that piece of knowledge, and as the sequences of mental states occurring inside peoples' heads get more varied and more representative of the external reality, the abstract representation of the totality of all conceptual constructs grows larger and more complex. This is the process of expansion of the abstract state space of conscious states, which, as we posited earlier, would eventually lead to the birth of a new, higher-dimensional universe.

The details of this process depend on the content of the conceptual constructs in abstract knowledge space, and in order to visualize it we need to have some idea of the relative proportions of the different topics that preoccupy people's minds at the present times. These details are roughly reflected in the categorization of the economic activity and the relative size of the categories measured in money, since most of people's daily activity is related to the economic output, with the rest being people's private lives where they typically consume the products of other people's labor. In terms of topics of interest in the free time, the relative proportions are roughly reflected in the media (mostly the electronic and printed ones), and the remainder of the mental activity is preoccupied for the most part with conversations with other people, typically consisting of storytelling from one's daily life.

The main categories of economic activity, which takes up about half of the conscious time of nearly all people living in our times, are agriculture, industry and services. They follow roughly our earlier distinction between manual and intellectual labor with different proportions of representation of physical material structures versus conscious states in other people's brains. The agriculture and industry sectors involve manual labor to a greater degree and account for about one third of economic output, while services, which may also comprise mostly of manual labor, but may have too a significant intellectual component, account for about two thirds of the economic output in terms of money.

This situation is a development from earlier stages of economic organization with higher proportions of manual labor and correspondingly larger agriculture and industry sectors. Nowadays, such higher proportions are characteristic for less developed economies, and the trend for all economies is to move to higher proportions of the services sectors and towards more intellectual occupations as manual labor is being replaced by automated machines and more sophisticated tools. This development corresponds to the shift in economically developed societies towards more abstract and more complex conceptual constructs, with greater emphasis on the importance of people's intellectual skills.

The other half of people's conscious lives—their private lives—involves similar proportions of representation of physical versus intellectual states. In more developed societies basic household activities (such as cleaning, cooking, tidying, etc.) are better organized and take up less time, which leaves more time for more intellectual types of activities, such as having conversations or watching television. The main topics of interest here are the political and social life, entertainment, sports, and the major events in the private lives of others, as reflected in the main topics in the media and in people's private conversations.

Politics and money are two of the most significant topics preoccupying people's minds, with politics being about the major decisions affecting the lives of the largest groupings of people and money being about the way people cooperate and exchange the products of their labor leading to a system of division of labor of increasing sophistication. These topics have high prominence in the media and in the shared knowledge of all people in a society, and therefore they deserve special attention, which will be duly paid to them in the next section. Together, they probably account for the largest chunk of shared knowledge among all people on Earth nowadays. The other highly significant chunk of our collective knowledge is the newly created one, at the forefront along the evolutionary trajectory of the process of accumulation of representations. This type of knowledge is for the great part captured by what we call science and the activity of scientists, and therefore it too will get special attention in a subsequent section.

Politics and Economics

Most societies nowadays are democracies, and in democracies the most immediate role of politics is to choose representatives of the people who will make the most general decisions about the social and economic life on their behalf. In a broader sense, politics is about the top-level decision making process on issues concerning the whole society. It has two distinct components—one is the true decision-making process on social and economic issues, and the other is a metaprocess about the way the political system works and who will get the power of decision-making.

The two main instruments the government has in its disposal for implementing policies are rules and regulations captured in laws and other mandatory decrees, and collection and redistribution of money through taxation and government spending. Thus, the majority of the decisions of the people making up the government are about money and rules of behavior, therefore politics and economics are intricately intertwined.

There are three major types of ideologies nowadays that dominate the political systems in most countries—these can be roughly denoted as capitalism, socialism, and religion. Capitalism and socialism take up the right and the left side of the political spectrum in developed countries, while religious political formations are more dominant in some developing countries, mainly in the Muslim world. Most of the political controversies are on issues that have different interpretation within each of these main ideologies and a different approach to solving the problems.

Capitalism and socialism are mainly about the role of government in society and in particular about the redistribution of money through taxation and government spending. The religious ideologies make decisions about rules and money as well, but the primary source of authority for them is not reason but the holy scriptures of ancient times. They are the oldest ideologies, rooted in tradition, while capitalism and socialism are relatively new ideologies which appeared on the political scene as secular alternatives to religion.

Capitalism is also the default ideology, postulating free market relations among members of the society, while socialism appeared somewhat later as a reaction to what the less affluent social groups saw as unjust mechanisms in the operation of the free market, which need to be corrected through redistribution of wealth enforced by the government. In modern days, socialists (or left-wing political forces) advocate in general more involvement of the government in the social and economic life and higher level of redistribution of income from the rich to the poor, while capitalists (or right-wing political forces) advocate the opposite—less involvement of the government and more free market relations (i.e., minimum redistribution of income). On that scale, the religious ideologies tend to be leaning to the left, but they also have some elements typical to the right-wing thinking.

In terms of relative size, or prevalence, of those three major types of ideologies, probably the most wide-spread is the default ideology—capitalism and the free market—since nearly all societies nowadays have free market relations as the basis of their economies and private ownership is the dominant type of ownership of assets and means of production. This, however, was not the case until about two decades ago, when the Soviet system of controlled economies tried to rival the capitalist system, and its failure was perceived as a triumph of capitalism which undermined to some degree the socialist ideology.

Socialism is the second major ideology, present in most countries where religious political formations do not play significant role, and, as we said, in religion-dominated societies the role of the socialist ideologies is played by the religious political forces who advocate greater involvement of the government in social and economic life and less freedom of choice for the citizens. Oddly, in Western societies the right-wing, conservative ideologies have more elements of religious thinking incorporated in them, while the left-wing, socialist ideologies are typically more progressive and more disconnected from religion. This is due to the tradition of the conservative ideologies trying to conserve the status quo that naturally emerges in free market relations, while the socialist ideologies oppose the tradition and try to correct the unequal distribution of wealth in the free market through government intervention.

However, although socialism seems to be a correction of some fundamental flaws in the free market system, it seems to underachieve in comparison with capitalism, at least on the coarse scale of historic processes unfolding over decades, as evidenced by the fall of the Eastern bloc. The reason for this is that more money available to the wealthier segment of the population (due to less redistribution of wealth) typically translates into more money available for investment in new economic enterprises and more innovation, which are the two key components of growth. So, in the long term a system which invests more can outperform a system which redistributes more, even at the lower end of the wealth scale, because as a society grows wealthier it becomes increasingly easier to aid the low income segment of the population.

This is the reason why we ought to consider capitalism the dominant ideology of our time, and also the most successful one, although this is only the long-term, coarse average, and in the short term in some individual cases a socialist system can outperform its capitalist alternative. This has lead to the familiar opposition between the two ideologies in the Western political systems, with interchanging success of each of them. The reason for the competition between the two systems and ways of thinking is that the government, and consequently the political and economic policies, of a country can be only one. In theory, we could imagine two systems existing side by side with people free to choose to which one they should belong to, but in practice this would mean that the wealthier segment of the population will choose less redistribution of wealth and the poorer segment more redistribution of wealth, bringing the level of redistribution to a minimum. Thus, in order to have a socialist system of political and economic governance the system should be a single one with compulsory rules of taxation for everyone. The only way to have both systems running and competing is if they alternate in time, and not in space, with left-wing, socialist governments and right-wing, capitalist governments taking turns. Given that the capitalist system seems to be more successful in the long run, this would mean that right-wing governments should have more turns than the left-wing ones, and since the terms are usually long-lasting (4 years in the norm) for practical reasons, it means that there must be frequent alternation between the two systems. Thus, the optimal form of governance seems to be two terms of conservative leadership and one term of socialist leadership taking turns, with possible occasional two-term spells of the socialist mode of government.

Besides the dimension of redistribution of wealth and involvement of the government in the social and economic life, there is one other important dimension on which societies differ, and this is the degree of freedom in the social and political life, as opposite of the degree of authoritarianism in the decision-making processes. In other words, some societies are more egalitarian in terms of authority and decision making, while others are more hierarchical. Typically, the more modern and more developed societies are of the former kind, while the less developed and more traditional societies are of the latter kind.

The historical process of the evolution of the contemporary political and economic systems from their emergence in the Middle Ages until today is a process of gradual deconstruction of authoritarianism and hierarchical organization and its replacement with distributed and more egalitarian social mechanisms of decision making. This direction of the evolutionary process is mainly due to the increasing level of collective knowledge and increasing sophistication of the individual knowledge of the members of modern societies. The continuous economic development over the last few centuries has led to an increasing level of complexity in the social and personal lives of people, making them increasingly difficult to be managed by a top-down, command approach and leading to dilution of power and, to some degree, wealth as well.

In this way, the less developed societies are again trailing back along the trajectory of this process and exhibit higher propensity for, and degree of, authoritarianism. Ironically, the socio-economic systems built on the purely socialist ideology were also highly authoritarian, with the idea to impose the ideology by force, thus jumping ahead in the evolutionary process directly to a stage of complete egalitarianism in social and economic terms. That, however, turned out to be an impossible leap, and the authoritarian approach used to enforce egalitarianism actually created economic stagnation and social rigidity typical for any authoritarian societies.

The primary reason for the failure of the authoritarian top-down, hierarchical system of decision-making is, as we said, the enormous complexity of the socio-economic structures and the added complication of its continuous transformation towards novel, more complex structures and relations. It is simply not possible for any human mind to conceptualize and to anticipate this development in order to be able to guide it towards success. We can think of the collective conceptual system of knowledge underpinning the social and economic phenomena as an entity in abstract state space forming a closed system, i.e., a self-contained set of entities and relations which develop out of their own internal dynamics without influence from outside. This is relatively true for the global society, since it is not constrained much by external factors, such as other societies or major calamities that would transform the way it operates.

In those terms, as a closed system, an authoritarian system would rely on mechanisms of learning that operate along the chains of authority—people in the lower ranks are expected to learn from people in the higher ranks, who in turn are responsible for instructing large numbers of subordinates in many aspects of their social functions. Such a system leads to low diversity of knowledge since it is impossible for one instructor to instill a large variety of knowledge in multiple pupils. In addition, the people in the higher ranks are usually concerned with more general decisions and in a command-style decision making process it is not clear how these should translate into the more detailed decisions needed for the operation of the lower ranks. All this leads to a slowly transforming, rigid socio-economic structure with relatively low diversity of knowledge. This used to be the case with the former Soviet-dominated socialist system of economies and the main reason for its collapse.

The Western societies, on the other hand, remained flexible by divesting authoritarianism and moving towards more freedom, including more free-market relations. This allowed a greater diversity of knowledge to cooperate and compete, i.e., to adapt to other knowledge, leading to a more complex society still functioning as one whole. This feat was not devised by a single person or a group of leaders, although leadership plays an important role in guiding the transformation processes. It was made possible by empowering the lower-rank individuals to make their own decisions and to find their own ways for increasing their personal wellbeing. This process pretty much resembles the biological process of natural selection and symbiosis among species, which we casted earlier as a manifestation of the more fundamental law of accumulation of representations. The evolution of the socio-economic systems seems to be explainable in similar terms as biological evolution, as an instance of a process of accumulation of representations.

The growing complexity of life in the developed economies is substantiated by a growing complexity in the knowledge systems of the individuals making up those societies. So, we can conclude that the driving force of the transformation process leading to economic growth is actually the increasing intelligence and sophistication of the economic actors, not the formal regulations typical for the type of socio-economic system governing the social life. This is a secondary phenomenon, an outcome of the relationships among individual actors. The higher level of knowledge and sophistication in developed societies is suited best by a system with more freedom and consequently more personal responsibility (since it is easier to taker care of your own life when you live in a wealthier society), and consequently capitalism is dominant in those societies. In poorer countries, people are tempted to prefer a command-style social organization where the government is required to 'care for the people', i.e., to take responsibility for the main aspects of their lives, such as providing food, housing, health care, education, etc., since they are too poor and unable to sustain themselves on their own. This allows authoritarian socio-economic systems, either socialist or religious, or a mixture of both, to flourish.

That is actually one of the reasons of the apparent failure of pure socialism, based on massive redistribution of income—it has been tried only in economically underdeveloped societies, since they are the ones to whom these ideas seemed natural. The redistribution of income, however, stifled investment and entrepreneurship at the individual level, and that led to economic stagnation and ultimately failure of the social system.

The capitalist system, on the other hand, incorporates more individual freedom and responsibility, leading to a higher level of knowledge, and in doing that it reflects better our earlier ideas regarding the meaning of life, the two categories of mental states (abstract vs. concrete), the ability to reduce everything to a single concept, the nature of right vs. wrong decisions, etc. The main reason why a system with more individual freedom and responsibility is more successful is that at the most fundamental level it corresponds better to the representational nature of matter and its current expression in human minds. As we discussed earlier, we can picture human brains as systems of highly concentrated representations, compressed enough to engender consciousness, while any other organization of matter forms only sparse representational structures. That picture of approximately equipotent concentrations of representations acting in a medium of rarified representations corresponds qualitatively to a social organization of approximately equal distribution of authority and responsibility, rather than a hierarchical structure with more authority at the top and less at the bottom. That is not to say, however, that personal wealth should be distributed equally as well, since it is not an expression of authority and responsibility, but rather of the relative quality and correctness of one's decisions, and these can differ substantially from one person to another.

Modern-day capitalism is far from an equal distribution of authority, but it is closer to it than socialism, and the development of the social relationships is going in that direction. The empowering of the lower social strata is made possible by the advancement of technology and the media, leading to higher involvement of people with various opinions (lobbyists, NGO's, etc.) in the decision-making process. This heightened degree of awareness of the major social issues in the general public in fact takes authority away from the top decision makers and leaves them less freedom of choice in the big decisions, as compared to earlier generations. Nowadays, when public opinion solidifies or shifts on some topic, it is pretty much clear how the government would react. It is expected to act in certain ways in response to public pressure, so it has less room for making independent decisions.

Unfortunately, what the general public wants is more often not a wise choice but demands for short-term gains which are detrimental to society in the long term. Mainly, the demands are for more money to be redistributed in favor of a certain social group, and more recently, asking to preserve the natural environment by stopping major infrastructure projects. The main problem with the demands for redistribution of money in favor of some social group is that people generally do not realize that money is simply a means of exchange of material goods and services, i.e., an abstract idea, and instead think of it as a limited resource which can automatically convert to goods and services. That confusion comes from the different logic that needs to be applied when thinking of money at the level of the individual, which is what people are familiar with, and at the level of the entire economy, which is a closed system with respect to money, and whose logic is unfamiliar to most people.

For example, at the level of the individual, increasing one's money possession is clearly a good thing and it translates readily into more goods and services, however, at the level of the economy as a whole, increasing the money supply does not translate automatically into a proportional increase in the availability of goods and services, but more likely into increased prices, i.e., inflation. This is precisely because money is a means of exchange, a measure of the worth of the goods and services produced by people, which in the short term is a relatively fixed quantity and cannot be influenced much through increases in money supply. In the long term the quantity and quality of the goods and services produced in an economy depends on the level of its sophistication and can grow exponentially if people build up on their earlier capabilities and achievements, leading to larger and larger discrepancies between more developed and less developed economies. Thus, the large differences between the levels of wealth of economies at different levels of development and of individuals within the economies are not an expression of the relative differences in general abilities and skills, which are fairly comparable among the majority of people, but an outcome of a long historical process of accumulation of small differences in the developmental trajectories of the compared entities, leading to a constantly widening gap in the relative levels of their development.

For this reason, a system which induces people to make consistently better decisions and put more effort and attention in their work can substantially outperform a system which is only slightly less successful in doing so. This is basically what happened in the competition between capitalism and the Soviet style socialism; the capitalist system radically outperformed the socialist one due to an accumulation of numerous small advantages over a long period of time. Historically, capitalism has shown that it can mobilize people to achieve more and more, leading to an exponential growth in the sophistication of the economy, while the socialist ideology has yet to show that it can do the same in a different way, based on its own notions of social justice and fair distribution of wealth.

In fact, at the level of society, wealth is not synonymous to the ability to acquire more goods and services for one's personal use, as it is the case at the level of the individual, but is an expression of the relative decision-making power of different individuals. This is so because, as we said, in the short term the amount of goods and services produced in an economy and available for redistribution is pretty much fixed. People with more wealth use better quality goods and services, but there are natural limits as to how much a single person can consume; one cannot eat much larger quantities of food than another one, or drive more than one car at a time, etc. What sets apart people with high amounts of wealth and people with little wealth is the ability to make decisions on behalf of other people. Wealthier people have most of their money invested, i.e., they are lending it to other people for certain purposes. Collectively, they decide the priorities in the development of the economy—what should be done more of, and what should be done less of. Through the decisions they make with regard to their investments they determine the developmental trajectory of the physical system of people transforming their environment on the surface of the Earth that constitutes the global economy. Physically, the decision making process is what matters and what determines the relative success of an economic system, and not the equality in the distribution of wealth or in the relative authority for decision making among members of the society.

From a systems science perspective, the most successful type of organization of a system consisting of comparable units is the so called 'small world' type of network, which can be represented as a hierarchy in terms of the degrees of connectedness of the nodes in the network. This type of organization is observed in many natural and artificial systems, including the brain, the internet, the biochemical reaction pathways in the human body, etc. Basically, it means that there are a small number of very important units in the system (nodes with high degree of connectedness), a somewhat larger number of somewhat smaller importance, etc., and the largest number of units has the least connections and therefore the least importance. The political system and the economy (including the individual businesses that make it up) are also such kinds of systems, and the reason for this is that this type of organization achieves optimal performance of the system.

This is one more reason why the capitalist system outperforms the socialist and the religious ones. A hierarchical distribution of wealth is natural, and it readily translates into a hierarchical organization of decision making, leading to optimal performance of the economy and exponential growth. Any attempts to artificially level the distribution of wealth result in suboptimal organization of the system and suboptimal performance, leading to stagnation of the economy in the long run.

The hierarchical organization of decision making translates into a hierarchical distribution of knowledge, since decision making is based on knowledge and one's knowledge is shaped by the types of decisions one needs to make. In particular, people with more decision making power near the top of the hierarchy need to possess more general knowledge, i.e., they need to know in detail the more general processes operating in the system, e.g., the inflationary propensity of the economy, or the competitive advantage of the labor force in one country versus another, etc. People with less decision making power near the bottom of the hierarchy need to know in detail specific facts about their job, e.g., how agricultural crops grow, or how to program a computer to do certain data manipulations, or how to operate a machine, etc.

This kind of distribution of knowledge also seems to be optimal from the perspective of our picture of the world as made up of representations, since representations too can be more encompassing or more specific, and a hierarchical distribution going from broader to narrower representations would cover all the possibilities and would leave no gap in the knowledge contained within the system. This organization allows for a smooth evolutionary process without any major conflicts or deficiencies, and this is generally what people like; major social conflicts are a necessary evil at best and they always involve traumatic experience for at least some part of the population.

The people near the top of the decision making hierarchy are people with more general knowledge, as we said,—these are senior level management, politicians, aristocracy (in some places), and otherwise wealthy people, such as athletes and entertainers. Their knowledge is more about principles; how things work in more general cases rather than in specific instances. There is, however, one more group of people who do not possess high levels of wealth but whose knowledge is also more general than the average, and these are the scientists.

In science there are some people as well who possess high levels of wealth and are near the top of the decision making process in society, but this is not the general case. Most scientists are middle class people whose job is to find out new knowledge, and precisely because of the nature of their job they need to posses more general knowledge in their area of specialization. However, that does not translate readily into a higher level of competency on the topics of concern for the society as a whole, such as economics and politics, unless this is precisely the area of specialization for the scientist.

This special group of people deserves more attention, since it is responsible for the creation of most of human knowledge (apart from areas like politics, economics, and medicine, where it is on par with the practitioners), spearheading the general evolutionary process of transformation of the material structures on the surface of the Earth, and it will be the subject of the next section.

The Role of Science

Scientists play a leading role in society, alongside with the political and economic elite, because they are the creators of new knowledge and their efforts 'push the envelope' of the evolutionary process of accumulation of representations, which in our times advances mainly in the direction of accumulation of new human knowledge and technology. Scientists achieve this in a more covert manner than people possessing decision-making power or wealth, through invention of new technologies that influence people's lives and through influences on decision makers by formulating theories which are put to use in practice. In this way, they determine the direction of the more general process of the evolution of human thought by defining the patterns of thought, or ideas, that people will be interested in over longer periods of time, say, decades, while the practitioners set the agenda for human thought on shorter time scales—in the present and in the immediate future.

To achieve their goal of discovering new knowledge scientists have developed a different culture of thinking than what is customary for the general public—one distinguished by high degrees of discipline and doubt at the same time. These two hallmarks of the scientific tradition work in opposite ways; the strict discipline of thought means that the people who are acquiring scientific knowledge follow the same paths as the originators of that knowledge, leading to more cohesion of ideas among the members of the scientific community, while the emphasis on doubt allows scientists to reject obsolete ideas and adopt new ones more easily.

This is in contrast to the large variety of ideas and opinions in the general public, which was described in earlier sections. Science is about finding out the truth, and since the truth is only one, it is natural for scientists to have greater cohesion of ideas. The problem with the greater agreement is that if it happens to occur away from the truth, i.e., the particular idea turns out to be wrong, it would be more difficult to change people's knowledge. This is where the tradition of doubt comes into play. A culture of doubt demanding high degree of certainty in the experimental verification of an idea means that scientists are well prepared to change their agreed position when faced with new evidence supporting a different explanation.

We can picture this state of affairs also using our abstraction of particles in state space. The higher level of agreement represents a tighter particle aggregation in abstract state space; we can picture it even as a single particle representing the single, agreed-upon idea, to which the overwhelming majority of scientists subscribe. In contrast, the opinions of the general public form a diffuse entity of particles with concentrations around the main points of loose agreement in those opinions. Such an entity is more active internally; it transforms and oscillates, but remains mostly in place and rarely exhibits coherent motion in one direction (i.e., when the public opinion shifts on some subject). The scientific ideas, on the other hand, are more stable concentrations of particles, which occasionally perform 'quantum jumps' from one place to another, corresponding to the subversion of the old paradigm and the endorsement of a new one.

The greater coherence around a single idea constitutes also a more definitive state than the loose agreement of diverse opinions. In the abstract state space of mental states it is a state with less energy but greater lumpiness of the particles representing mental states, due to the greater probability of finding the particle in one particular state. All this in effect indicates a more mature stage in the evolutionary process of matter towards the creation of a new universe, which is why science should be considered at the forefront of the evolutionary process, setting an example for the rest of society of the way people will think and act in the not so distant future.

Science is also less hierarchical than the typical business or governmental structure, with authority more widely distributed among the members of the community. Unlike business and government, where people have fixed roles and it is clear who is subordinate to whom in the decision-making process, in science the authority resides with the person who makes the most persuasive arguments, and at different times these could be different people. Decisions are taken in the context of the peer-review process, where several peers evaluate the claims of a scientist regarding the proposition of new facts or hypotheses, and collectively decide on the merits of the proposition. In this way a single individual does not play a fixed role in the decision making process by giving instructions to subordinates and receiving instructions from superiors, but the role of the authority instructing others how and what to do is taken in turn by whoever happens to be the most knowledgeable person in the given circumstances. This organizational trait of science also seems to be ahead of the rest of the society, showing the type of organization it will move to in the future.

The more advanced, or mature, stage of development of science can be understood also in terms of representations as the fundamental constituents of our picture of the world. Earlier we saw that the general tendency of the evolutionary process of the universe is towards a state of greater representation, which we expressed also as a law of accumulation of representations. Human consciousness, as a novel type of organization of matter leading to a higher dimensionality of material structures, also strives towards greater representation of the external world, and the best way to achieve this is by forming mental states, or knowledge, that is as veridical as possible (at the current stage of development) of the external world. This is exactly what scientists try to do professionally, albeit for a small subset of the phenomenology of the external world, due to their narrow specialization.

Scientists form their knowledge through experiments, i.e., through observations of controlled and simplified situations, where they try to figure out a single aspect of the complex phenomenology of the real world. In this way they try to construct simplified models (i.e., representations) of the more complex phenomenology, which in certain circumstances function in the same way as their real-world counterparts. This approach is in fact a way of creating representations, both inside the minds of the scientists and in the outside world in the form of the experimental apparatus (including the physical realization of mathematical and computational models) which represents a simplified version of the real-world system under study (the computational models are physical representations of the modeled physical system because they are physically instantiated by flows of electrons along wires; thus, individual events in this physical process, e.g., passing or stopping current flow, correspond to individual events in the studied system, e.g., a constituent unit contributing or not contributing to a measured property).

The most widely used tool in science for modeling the outside world is mathematics. It is a set of abstract relations based on the idea of a number, i.e., a comparison of relative magnitudes of analogous properties of different entities. Although mathematics itself is defined only in abstract terms, it maps readily to anything that has magnitude, and in this way serves as a universal modeling tool for real physical objects and phenomena. This is especially true in physics, where the behavior of the systems under study is highly regular and reproducible, rendering models an almost perfect match for the real systems and the physical laws formulated through mathematical equations the status of universal truths.

As any abstraction, however, mathematics does not have a reality on its own, i.e., it is not a physically existing entity, and the numbers (or information) cannot be thought of as the most fundamental building blocks of reality, but this role should be delegated to a concept that has some physicality to it, such as representation. Numbers and mathematical operations can only serve to model the workings of representations, but when we think of the world using our most fundamental knowledge, we need to imagine it as material, not as made up of mathematical abstractions.

The abstract nature of numbers is an advantage when it comes to building models of entities that behave in the same way in analogous circumstances, since numbers and the mathematical operations performed on them can represent the functioning of any mechanism through modeling only the relations among its constituents, without taking into account their physical nature (e.g., whether they are large aggregations of particles or individual particles). Thus, by expressing only relations without ontological identity mathematics succeeds in becoming a universal scientific tool, but it loses its ability to serve as the base for the ontology in the description of the world.

More specifically, the basic arithmetic operations of addition, subtraction, multiplication, and division represent confluent or opposing action and full interaction between the constituent elements of two entities in coherent or discordant way, respectively, which map onto the logical possibilities of relationships between interacting entities and in this way make it possible for mathematics to describe the phenomenology of any physical system, as long as there is some regularity in it.

This works even for individual particles, which from the point of view of the traditional physics ontology are point-like objects and therefore cannot possess any internal structure. In our ontology of representations the operation of addition would mean two representational entities acting together on another one, while multiplication would mean full interaction between two representational entities (just like the multiplication of two numbers can be depicted as each unit in one of the numbers engendering the number of units attributable to the second number—a form of full interaction between their constituent units). The operation of raising to the power of 2 would model a self-interaction, i.e., an interaction of a representation with an entity identical to itself, while higher powers would indicate multiple instances of such interaction (e.g., three-way or four-way, etc.). The opposite operations of subtraction, division, and taking a root would carry analogous meaning but in an opposite direction, i.e., expressing a mismatch or increasing the degree of mismatch of two representations.

In summary, we can say that the mathematical operations need to be understood as models, i.e., another, indirect way of thinking about reality, rather than reality itself. They are very helpful for scientists, especially in physics, but they remain only models, and in order to understand the real world we need other concepts that match reality directly, rather than only the relational aspects of it.

That is also the reason why the ontology of modern physics cannot serve as the most fundamental description of reality, but this role has to be taken up by a philosophical ontology. The thinking of physicists is rooted in mathematics, which is really useful in practical terms—for calculating and predicting the highly regular behavior of inanimate physical systems, but the mathematical description itself does not have the capacity to provide an ontology for the description of the world, and this needs to be done using the conceptual apparatus of all human experience, not just mathematics.

Science is one of the most important human activities, since it creates most of our new knowledge, and together with the political and business elite it shapes our immediate future. The next few decades, just like the more distant future, can be predicted in general terms if we know the direction of development of science and politics, and we will attempt to do just that in the next section.

The Near Future (The Next 50 Years)

In the preceding sections of Chapter 4 we outlined some of the outstanding characteristics of the present state of development of our knowledge of the world in light of the representational ontology developed in the preceding chapters. The trends we identified will likely persist in the next 50 years and shape the developments throughout this period.

We formulated the meaning of life at the present moment as the process of transforming our immediate environment on the surface of the Earth, including people's minds as physical entities. This is the most general formulation of the trend in the development of our civilization, and it will continue to be valid 50 years on and even much longer than that. It is a process of gaining knowledge, or accumulation of representations in the most general sense, and this knowledge is captured in particle structures inside and outside the human brains. This means that the complexity of these particle structures will grow continuously, and 50 years from now there will be many new types of structures that are unimaginable now—new technologies, new ways of moving physical stuff and information, new cultures and ways of thinking. Still, 50 years is not such a long period that the world could change beyond recognition; it is still within the human lifespan and the changes will be comprehensible by today's standards.

In terms of the way we think and perceive the world, i.e., our conceptual structures, we said that it is best to think only in physical terms and avoid abstractions. That demand, however, is unlikely to be fulfilled in the next 50 years, since the current trend is towards expansion of knowledge and generation of new ideas, which prevails over the consolidation and reconciliation of existing ideas. Therefore, although many of the current controversies in the scientific and social life may be resolved, many new ones will arise, and overall the level of debate and ideological opposition is not likely to subside, although it may shift to new modes of argument. Thus, the goal of being able to reduce all of one's knowledge to a single concept may be attained only by a small community of people, but the majority are likely to continue the inertia of traditional modes of thinking and belief systems.

We focused on the judgments of right and wrong (as well as good and evil) as the most important singular mental events. This type of judgments will continue to play an increasingly important role in our lives, as our activities become more structured and our decisions make more and more impact on the system as a whole. Our societies are becoming more interrelated and integrated, progressing towards a single global community of people, the first signs of which may be evident 50 years into the future. This trend means that people's activities will become more organized and more planned in advance, which would make decision-making even more important for the way the global community system functions. This is the case with more coordinated structures, like living organisms, where every unit needs to play its role in order for the whole to function properly.

The advantage of a more structured and more coordinated life is that it is easier to achieve one's goals, and subsequently it is a happier life, but the downside is that diversity goes down, as well as the individual freedom of choice, i.e., a higher degree of organization requires also a higher degree of conformity from the individual. There will be more rules in the near future and more peer pressure to obey them, but this also requires more knowledge on the part of the individual in order to function properly, which is in fact the main determinant of progress in line with the fundamental evolutionary process in the universe.

The increased level of coordination will have the other beneficial consequence of reducing conflict and disagreement. Although it is unlikely that 50 years from now all major ideologies would have converged to a single one, the conflict and disagreement would have shifted to softer and less violent modes. Physical violence, most notably wars, could be a thing of the past, and conflicting viewpoints would be more readily resolved through logical reasoning based on inferences from more basic principles.

We also noted earlier that in terms of the two main types of conceptual representations—those of the physical environment and of abstract notions, including other people's minds—there is a shift towards higher levels of abstraction with the economic and social development of a society. This trend looks set to continue in the next 50 years sustained by the rapid economic development of the poorer countries. In order to achieve higher standards of living, people will need to learn to perform more complex tasks, and that entails more complex conceptual structures and higher level of abstractions. The leading role, still, would belong to the economically more developed communities who will move to a state of even more complex knowledge and even greater abstractions when many of today's conceptual constructs which are the purview of science at the moment enter the public domain.

The major change in our understanding of the world, which might be even revolutionary in its effect, will come from the better understanding of other minds and the functioning of the human body. Science is gradually unraveling the mechanistic workings of the human brain, and if consciousness gets explained—along the lines of the argument in this book or in some other way—then scientists will start viewing the human mind in a different way and the new worldview may creep in into the public domain. This new knowledge can have a major impact on people's views on free will, their main purpose in life, one's responsibilities towards others, etc. It will change also the amounts of time people spend on various activities; for example, people may devote more of their spare time to studying and learning scientific knowledge at the expense of entertainment and religious practice.

Although most people are unlikely to learn quantum mechanics and to learn to imagine matter in terms of its quantum phenomenology, some aspects of the logic of quantum processes may make inroads into popular culture, especially when related to the workings of the economy or the human brain and body in relation to health, for example. These systems are constituted of entities with high density of representational relations among their constituents (human minds in the case of the economy, and biochemical mechanisms in the case of the human body), and as we saw in the preceding chapters, these representational relations are of the same type as the representational relations between particles and therefore obey the logic of quantum processes.

The introduction of quantum logic in public thinking will be further motivated by another development in our technology—the advent of the quantum computer. Although it will be restricted to performing only some very specific types of operations initially, it will boost the interest in the workings of the quantum aspects of matter and that in turn will improve people's understanding of the world by bringing their conceptual framework closer to reality.

The new fundamental picture of reality will have wide-reaching implications for people's culture, everyday thinking, and the political and social life. The addition of quantum logic to classical logic will enhance the representational veridicality of people's conceptual frameworks, thus reducing the number of mistakes people make and leading to a greater coherence and agreement in their opinions. It will allow for a better functioning of the economic and social mechanisms, and consequently for greater productivity and capabilities of transforming the physical environment. These enhanced capabilities will be needed urgently, since humanity is coming closer to the limits of natural resources that can be harvested from the surface of the Earth and better technologies for recycling and more efficient use will be required to maintain our living standards.

It may seem that humanity is faced with many environmental and capacity challenges in the immediate future, but they all boil down to one thing—energy. Since all natural resources are just different arrangements of atoms in molecules, and the atoms merely get rearranged in the chemical and mechanical transformations, but not destroyed, the key to solving all of our resource problems is finding enough energy to rearrange the atomic structures the way we want. Currently, we are relying mostly on fossil fuels to do that, and in the near future we will be trying to replace the carbon-based fuels with cleaner, renewable energy sources, but there is one resource that can provide practically unlimited and safe energy supply for a very long time and resolve all our energy problems, and this is nuclear fusion—the same process that powers the Sun and the other stars. Although we still haven't mastered this process to make it energy-positive (i.e., producing more energy than it consumes), and it is not certain that we will ever succeed, it seems plausible that it may be mastered within the next 50 years and that will result in a totally new world order in terms of energy and the technologies we use. It would even make possible the sequestration of greenhouse gases from the atmosphere, thus reversing the negative impact of human activity on the climate.

There is one more technological development which may be realized within the next 50 years with the potential to transform our lives completely, and that is the advent of the quantum computer, as already mentioned. This development will not occur in a single step at a specific point in time, but it will be a gradual process of accumulation of technological advances leading to more and more advanced computational capabilities. In fact, we already have simple, small-scale quantum computers that can perform basic operations on a few qubits. The challenge is to scale it up to larger numbers of qubits and computationally more complex operations. The real power and advantages of quantum computation will become evident when this technology starts to get integrated with classical computers and their logic of operation. The quantum computers will allow us to model and control physical processes at the level of individual atoms, leading initially to advances in nanotechnology, biochemistry and inorganic chemistry, and solid state physics, enabling us to produce a vast range of new materials and to completely recycle everything that we use. The main advancement, however, will be the ability to model consciousness and to construct artificial conscious systems. Although this may not happen within the next 50 years, and human-like artificial consciousness is probably centuries away, the foundations for this achievement will be laid by the development of the quantum computer and its integration with classical computers.

The reason why quantum computers are a necessary ingredient for simulating consciousness is that they possess a property which we may call 'situational awareness', something that is lacking in classical computation. Classical computers can simulate logical steps and decisions based on comparison of two states, but quantum computers can simulate decisions made by weighing multiple possibilities simultaneously, just like we do in our normal conscious activity. This is a fundamental difference stemming from the different levels at which the physical systems instantiating the two classes of computational devices are realized. Quantum computers are realized at the level of individual particles, thus exploiting their representational nature, while classical computers are realized at the level of aggregations of particles where the representational nature of matter gives rise to reliable cause and effect relations that serve as the basis of the computational process but the representational properties of individual particles cannot be exploited for computational purposes, and in fact they lead to undesirable side effects that are considered 'noise' or 'heat' and impede the classical computational process.

The representational nature of matter itself, as we described it in Chapter 1, is responsible for the more powerful computational capacity of the quantum computer, and in that way makes it a more fundamental computational device than the classical computer. We can think of the classical computer as a limited version of a universal quantum computer—one lacking the 'situational awareness' property, or the ability to perform computations in parallel, which is an inherent property of particles of matter.

That same property is also the physical substrate of what we call 'consciousness', as we saw in Chapter 2. So, in order to model consciousness in an artificial system we have to include the 'situational awareness' property of matter as a property of the system we create, and that cannot be done with classical computational methods since the number of calculations needed to simulate the quantum behavior of even a single particle is staggering. With quantum computation the system simulates itself, i.e., we use a quantum particle to simulate the quantum behavior of one or many entangled quantum particles, which is perfectly doable.

The role of quantum computers in the development of artificial consciousness is currently not appreciated, mainly because it is unclear to most scholars what consciousness is. In light of our hypothesis regarding the representational nature of matter, however, quantum computation is the key component of an artificial conscious system and its technological development should be prioritized if we want to speed up the pace of technological progress. The optimum policy for the next 50 years is probably to address all the immediate issues related to the environment and the limits of natural resources, and to channel the remaining available capacity into the development of artificial conscious systems based on quantum computational devices. This, however, cannot happen before the nature of consciousness is demystified and widely understood by the majority of people. If that happens, it has the potential to transform all of our cultural, social, and economic life and to make life better and easier in general for virtually all people.

The technological advances, and the increased level of knowledge about the world in general, will have an impact on the social and economic life and relationships between individuals. As we said, technological developments are the answer to the constraints of scarce resources and climate change, putting pressure on people to become more knowledgeable and aware of the way society and industry work. The increased level of knowledge will lead to greater agreement on what is well known, but it will also expand the horizon of existing knowledge, leading to greater disagreement on what is vaguely known. However, this is a positive development which will allow for more complex policies and processes to be implemented by the government and private businesses, leading to greater prosperity and ability to transform the environment according to human needs.

The result of the increased prosperity will be a shift towards the default ideology—capitalism—or rather, a transformed version of capitalism, since it is an ideology that constantly reinvents itself and adapts to the new social order. Socialism will lose ground, although it will transform and adapt too, since its basic tenet—the opposition between the lower and the upper strata of society—will lose its relevance due to the raising affluence of the society as a whole. In the near future people will think of themselves less in terms of being rich and poor and more in terms of how they fit in a highly diverse society, of their role and contribution to social life.

The new understanding of the social, political and economic issues will be founded on a better understanding of the concept of money and its workings on the scale of the whole economy and across economies. This will make people more likely to move away from the socialist reasoning schemes about the optimal social and economic mechanisms, which will be exposed as internally inconsistent, and towards the capitalist ones, which place more responsibility and decision-making power on the individual instead of the government.

Paradoxically, the shift towards capitalism and away from socialism will result in a more egalitarian social structure in terms of decision-making power and responsibilities, and the reason for this will be the increased level of knowledge of the individuals that make up the fabric of the society and the economy. More knowledgeable individuals means more complex social relations and mechanisms, which would make it impossible for society to operate in the frameworks of the old hierarchical structures. New ways of governing and management will emerge, with more independence of the subordinates in their decision making and choice of action. This may lead to more egalitarian income distribution as well, although this may not necessarily be the case. Anyway, income inequality will become less and less of an issue in the near future.

What people in the future will be concerned with is not that much how much money and personal possessions they have, but more how much knowledge and influence they have. People will be rich enough to have the basics needed for subsistence secured, so they can go beyond mere survival and focus on more long-term goals and more fundamental questions. This will change their culture and the political and economic organization of the societies. Initially these changes will emerge in the developed countries, but with the developing ones rapidly closing in, the boundaries between cultures and states will become smeared and towards the end of the 50 years period a borderless, global society of all people will start to take shape.

People's personalities will change too, shifting towards a more contemplative, more rational style of thinking and behaving, akin to today's personalities of people in science or in the political and business elite. This type of personality will become the norm and it will be shaped by the increased level of knowledge resulting from the stronger emphasis on education, which will not cease with the end of the formal schooling stage of life but will continue throughout the productive life of the individual. People will come to take life and their roles in society more seriously, paying more attention to the consequences of their actions and the decisions they make. These new attitudes will be the basis of the new political and economic organization with more individual freedom and responsibility and less hierarchical decision-making.

After reviewing the consequences of the new picture of matter and consciousness based on the concept of representation proposed in the initial chapters of this book we can finally draw some conclusions about the way this new worldview should impact our own personal lives and habits of thinking. Our elaboration of the possible developments over the next 50 years will help because this is the general direction of the developmental trajectory of our civilization, and we just need to follow it in order to understand how it will impact out personal lives.

Conclusion
(What Does It All Mean From a Personal Perspective)

So far we developed a new paradigm for understanding the world that we live in based on the idea that all our understanding can be reduced to a single concept—that of representation—and everything else that we can think of can be seen as different manifestations of that most fundamental concept. We also discussed the consequences of this worldview for our collective thinking and the possible implications it may have for the near future. Undoubtedly, the realization that we live in a world made up entirely of representations ought to have implications for our personal, day-to-day thinking and behavior as well.

The most obvious and immediate implication for anyone who adopts this worldview would be the persisting shift of attention towards the more fundamental questions and problems in our contemporary thought. Of course, most of the time nearly all people in our times are preoccupied with practical issues from their daily lives, and this cannot change (and should not change) substantially, but the understanding of the idea of the fundamentality of the representation concept would gradually make inroads into the daily thinking and lead to realizations of how different phenomena and conceptualizations from daily life can be reinterpreted in terms of the fundamental picture of reality. This is a very slow and gradual process, given the long history of disengagement of the general public from the profound philosophical questions, and it will take generations on average for the attention to shift towards those questions in the public mind.

On the personal level, acquiring understanding of the more profound philosophical questions is gratifying, but more importantly, it brings one closer to the truth with the implication of being able to think and act more successfully. Just like with particles, a person with a better representational picture of reality will be able to better navigate the conceptual space of other people, experiencing fewer and less violent collisions with their viewpoints, and being able to act on and transform those viewpoints more successfully. It should be noted that better understanding necessarily leads to higher ability to influence other people; it does not depend on the compliance of others on average.

However, the ability to influence other people should not be a goal in itself, and if it happens it would be a short-lived phenomenon until most other people adopt the fundamental understanding of reality, enabling them to make better decisions as well. The main goal in one's life should be to promote the basic evolutionary process in the universe, as we pointed out in the discussion of the meaning of human life, which would bring one closer to the truth (i.e., form a more veridical representation of the external environment in terms of one's conceptual structure), and the ability to persuade and lead others who have not achieved this ability yet is just a temporary side effect.

In terms of one's conceptual structure, the ultimate goal would be to express everything in terms of the most fundamental concept—that of representation, i.e., to see and think of everything as different manifestations of representational interactions. This goal may seem a bit outlandish from the point of view of present-day common sense, but it is probably not that difficult to achieve. It only needs the development of a habit to always return to the idea of representation and try to understand other concepts through it in a multi-layer hierarchical framework of successively more complex concepts. The main impediment to achieving this goal is, however, our limited knowledge of the world. We still do not have complete understanding even of man-made phenomena, such as the economy or individual reasoning patterns in common situations, and the ability to express all knowledge via representations will be achieved only after we develop proper artificial intelligence, which presupposes a complete understanding of human thinking in common situations, apart from the creativity component of insights and novel reasoning constructs.

To achieve the goal of acquiring more knowledge one needs to focus on the ideas of right and wrong, as described in the earlier sections. More knowledge, in fact, means more correct knowledge, not just any kind of knowledge. Correct knowledge is what brings one closer to the truth, constituting a more veridical representation of reality, and it is the physical instantiation of the idea of 'good,' as opposed to 'bad,' which is instantiated by incorrect knowledge.

To that end, one needs to pay more attention to one's personal actions and decisions and their significance in the context of the totality of all matter understood as representations. That type of action requires more knowledge, but it also has the effect of producing more correct knowledge in the minds of the people who practice it. Increased awareness of one's own actions and their consequences goes hand in hand with having more correct knowledge and being closer to the truth, and the causal relations between these two phenomena go both ways.

It is possible to understand this process also in the framework of the representational picture of reality developed thus far. We conceived of the brain as the densest aggregation of representations in the universe. That led to our picture of free will as the unrepresented part of the brain structures, where new knowledge is created. This occurs mainly during the moments of learning and insights and is boosted by paying attention to a particular question. That unrepresented part, most of which resides within the shared entangled state of consciously organized matter, is what drives the evolutionary process of matter in the universe and defines its direction. So, one's personal efforts and decisions regarding what to attend to and what to consider right or wrong turn out to be the defining factor for the future of the entire universe. It makes the difference between the universe being well reconciled and smoothly evolving or being unevenly developed and violent. We can see this effect on our human scale of our societies in the present time—there is relatively little hostility and violence within societies, where people have come to share most of their knowledge and culture, but there are still some tensions across cultures and societies, especially across religious ideologies, which are logically irreconcilable. It would take time for these differences to get reconciled and to achieve permanent peace and prosperity, and how soon that will happen and in what way depends on everyone's individual efforts, on their free will, i.e., the unrepresented conscious structures inside people's brains.

This idea may be extrapolated into the far future of the universe, leading to the conclusion that the level of reconciliation today determines the level of reconciliation, and therefore violent interactions, in the future when the universe will start engendering four-dimensional material structures. Reconciliation does not mean only what we currently understand as physical violence—wars, crime, etc.—but also ideological reconciliation, i.e., agreement on the right ideas. With rising affluence we are sure to reach a point when using physical violence to obtain limited resources will be a thing of the past, but there is no end in sight for the ideological differences of opinions and reasoning constructs, and this is actually what needs to be achieved in order to ensure a peaceful future indefinitely.

Thus, the conclusion from this entire story for one's personal behavior and action is to be knowledgeable, which is best achieved by focusing on what is right and what is wrong, and to pay better attention to the more fundamental questions in one's knowledge framework. Besides everything else that we understand as being a good citizen, this extra ingredient has the potential to make the world a better place.

www.ingramcontent.com/pod-product-compliance
Lightning Source LLC
Chambersburg PA
CBHW031833170526
45157CB00001B/281